A Light History
of Hot Air

A Light History of Hot Air

Peter Doherty

MELBOURNE
UNIVERSITY
PRESS

MELBOURNE UNIVERSITY PRESS
An imprint of Melbourne University Publishing Ltd
187 Grattan Street, Carlton, Victoria 3053, Australia
mup-info@unimelb.edu.au
www.mup.com.au

First published 2007
Text © Peter Doherty 2007
Design and typography © Melbourne University Publishing Ltd 2007

Edited by Sally Moss
Cover and text design by Peter Long
Typeset in 10pt Baskerville by Midland Typesetters, Australia
Printed in Australia by Griffin Press

National Library of Australia
Cataloguing-in-Publication entry

Doherty, Peter Charles, 1940– .
    A light history of hot air.

    Bibliography.
    Includes index.
    ISBN 9780522854077 (hbk.).

    1. History–Miscellanea. 2. Science–Miscellanea.
    3. Science–Popular works. 4. Climatic changes–Miscellanea.
    I. Title.

902

This project has been assisted by the Australian Government through
the Australia Council, its principal arts funding and advisory body.

Contents

Introduction   vii

Floating in Air   1
Alphabet Soup   15
Life, Gas and Hydrocarbons   25
Soaring with Eagles   36
Burnt by the Sun   48
Iron Horses and Balladeers   71
Hearth and Home   97
The Iceman Cometh   110
Night Lights   124
Imagining the Red Baron   137
Beacons   162
Tall Ships, Black Gangs, 'Bully' Wars   176
Firefighters   197
The Hot Air Diet   210
Becks and *Bleak House*   235
Political Hot Air   241
Flying the Concorde   251
Heating the Planet   256

Notes and Selected References   274
Abbreviations, Terminology and Conversions   285
Acknowledgements   290
Index   291

Introduction

Hot air is such a vast topic that a
single book by one person could not possibly be authoritative. My light account is meant to intrigue, entertain and
even provoke a little, rather than to be exhaustive or comprehensive. Some of the stories and insights are based on the
ordinary, everyday experiences of my wife Penny and me as
we lived, worked and raised children on three continents:
Australia (Brisbane, Canberra and Melbourne); Scotland
(Edinburgh); and the USA (Philadelphia and Memphis).
Grandparents and parents appear, partly because much of
what is discussed here happened in their lifetimes.

By definition the perspective of a history such as this
derives more from the past than from the present; under-
standing the evolution of events, technologies and
perceptions benefits from the distance of time. We in the
twenty-first century are experiencing such momentous

changes that, with the exception of the greenhouse gas story that is developing so rapidly and can't be ignored in a book on hot air, I've skimmed over many current happenings. There isn't much here about motorcars, though they are a major source of carbon monoxide and pollution. We all know about automobiles; there are many books about them and there isn't a lot that's interesting to say. On the other hand, the theme of climate change is a fairly constant subtext.

Also central to this story of hot air are the themes of burning and illumination. Both in the actual and the metaphorical sense, burning and illumination are irrevocably linked through the long march of humanity. Over the ages, we've lit big fires and gentle flames to open our minds, to warn of danger, to brighten our way through the darkness and to allow us to read in bed at night. As every photographer and cinematographer knows, the way a scene is lit determines how the subject is perceived.

Big topics such as global warming, environmental degradation, the cause of wars and the nature of political upheavals are juxtaposed here with modest, domestic themes: cooking, air-conditioning, sunburn, going to school on a steam train. Shining an indirect, tangential light has the potential to illuminate issues in new and unexpected ways against the backdrop of history.

Floating in Air

On zoos and the Serengeti

An easy stroll across Royal Park takes us from our narrow, two-storey Victorian house to the entrance of Melbourne Zoo, a fine institution that has its beginnings in the sixth and seventh decades of the nineteenth century. Around the mid-point of that walk stands a stark pile of stones marking the assembly point of the ill-fated Burke and Wills expedition which, with great expectations, set out in August 1860 to cross the Australian continent from south to north.

Apart from the nineteen humans from different regions of the globe, the most exotic creatures in the Burke and Wills menagerie were twenty-seven camels imported especially from India. After slogging across increasingly harsh landscapes, the explorers reached the distant, inland oasis of Cooper's Creek, where they made their sixty-fifth camp. Then, in the full heat of a Central Australian December, Robert O'Hara Burke and

William John Wills led a party of four in a dash north to the Gulf of Carpentaria. Only one member of that small group, James King, survived. By the nature of their deaths and the idiosyncrasies of Burke's character, they became, like Gallipoli and Ned Kelly, enduring icons in the Australian legend. Several of the paintings in Sidney Nolan's 'Burke and Wills' series isolate the two men and their camels in the stark, bare redness of the desert landscape. Others illustrate their solitary deaths on the banks of Cooper's Creek. The more effective and better organised John McDougall Stuart, who completed the south-to-north trip in 1862 and set the route for the later overland telegraph line, is much less well remembered.

Coastal Melbourne at the southern tip of the continent is a considerably gentler environment than inland Australia. On occasional summer weekends, when our bedroom windows are open and the clock radio isn't set to rouse us with classical music, the first thing we hear is the sound of the zoo monkeys chattering on the other side of Royal Park. Sometimes when the wind is blowing steadily from Bass Strait in the other direction, we wake suddenly to what sounds like a lion roaring right over our heads. Only it isn't a lion, but the gas burners of a hot air balloon floating across the city from the Melbourne Cricket Ground, rapidly losing altitude as it comes in to land.

At times, the balloons just miss the highest point of the roof above us, the now ornamental chimney pots. The pilot activates the burners, the flames leap high into the canopy, and the balloon lifts a little so that it can also clear the row of tall houses opposite. We are relieved as we recollect what it cost recently to replace the 1873 Welsh slates with a more

recent Spanish variety. On one of the balloonist–
entrepreneurs' busier mornings, as many as six or eight of
these exotic, brightly coloured visions, with their advertising
logos, skilled drivers and paying passengers in suspended
baskets, float across at different heights, then drop rapidly
out of sight as they settle onto the flat grassland of Royal
Park. These early morning venturers are then whisked off to
a champagne breakfast in a city hotel, while the balloon crew
deflates the canopy and packs everything onto the back of a
truck. Watching this wind-up phase as we stride around the
walking circuit in the park takes us back to our own experience of ballooning in Africa.

Also less hostile than the Australian interior crossed by
Burke and Wills is Kenya's Mara landscape. The vegetation
is different and the animal inhabitants are more varied, and
in some cases considerably more dangerous, than emus
and kangaroos. Setting out from somewhere near the luxu-
rious, tented Governor's Camp, our floating chariot drifted
above the annual wildebeest migration in the Serengeti
region of Kenya and Tanzania. Before we landed—a process
that involved the wicker basket dragging across the ground
on its side till the canopy above us fell to earth—we were told
to brace and not jump out prematurely. Some of the big cats
might be hungry and ready to pounce on a slow-moving,
poorly protected, inferior creature that suddenly dropped in
at breakfast time.

The balloon ride was pure magic—quite different from
the other African safari adventure of bouncing across the
ground in a four-wheel-drive troop carrier as the guide tries

3

to manoeuvre close to a group of elephants or, if you're very lucky, a leopard or a rhino. The animals seemed to ignore the canopy floating above them and were completely unfazed by the sporadic roar of the gas burners. We looked down on thousands of wildebeest, crocodiles sunning themselves on muddy banks, hyenas, wart hogs, giraffes and different types of gazelles. A small herd of zebra grazed contentedly while a pack of lions ate one of their species just off to their left: why worry if the lions are well fed? Much of what wild animals do with their time is concerned with maintaining their calorie intake. What I remember most is a great sense of peace and the extraordinary variety and vibrancy of the spectacle that stretched beneath us. On our way to a picnic breakfast of bacon and eggs cooked over the multi-use balloon burners, we enjoyed a vision that no human born more than 230 years ago could have known.

Walking across Royal Park, past the Burke and Wills monument to visit the exotic animals at Melbourne Zoo, allows us to relive just a little of that African safari experience. But I can never just pass the cairn that commemorates the two explorers without recalling their tragic fate. Having lived much of my life in the United States and Australia, what fascinates me about Burke and Wills is the contrast between their experience and that of Meriwether Lewis and William Clark in their earlier (1804–06) east–west crossing of North America. While they encountered many difficulties, Lewis and Clark travelled much of the way by water and survived to tell the tale. Burke and Wills marched into increasing dryness, though they died on the banks of an isolated stream.

It's not inconceivable that Burke and Wills could have considered packing a hot air balloon when they set out on their 2600-kilometre odyssey. As it was, besides its multiple camels and humans, the expedition manifest listed twenty-three horses and some twenty tons of equipment, including a bathtub and cedar and oak dining tables and chairs. All this was loaded onto six wagons, one of which was designed so that it could be converted into a boat if they did indeed find the much-desired but regrettably non-existent inland sea. Although marine fossils are found in the dry heart of Australia, the salt waters receded around 100 million years ago.

A hot air balloon would certainly have been of more value to Burke and Wills than a dining suite or even a boat, as it would have allowed them to see far ahead on those flat, dry plains, then across the marshes that prevented them from reaching the shores of the Gulf of Carpentaria. Yet that scenario belongs in a Hollywood-type fantasy or a book for small children; by the time Burke and Wills reached Cooper's Creek they had dumped most of their equipment and could barely drag themselves across the terrain, let alone lug a balloon. The idea of balloon exploration in 1860 is not, however, outrageous, as the basic technology had been around for more than seventy years. The Union forces deployed four reconnaissance balloons in the 1861–63 stages of the American Civil War, and some use was also made of them for cartography during the nineteenth century.

On 18 September 1783, the brothers Joseph and Jacques Montgolfier flew a paper-and-cloth hot air balloon in the presence of Louis XVI, Marie Antoinette and 130 000 other people. The passengers were a sheep, a rooster and a duck, with the first human 'aeronauts' ascending in October of that same year. The Montgolfiers achieved the reverse of the Icarus myth. Icarus flew too close to the sun, then fell to his death when the solar heat melted the wax that attached his wings. A hot air balloon flies because the heat from the burning gas, or wood and straw in the time of the Montgolfiers, causes the molecules of air to move further apart. Once the flame is extinguished, the balloon settles gently back to earth. That made balloons rather dangerous in the days before propane tanks, as, apart from the risk of fire, they could easily run out of fuel and dump their human cargo in very inconvenient places, such as rooftops, with fatal consequences.

At about the same time that the Montgolfiers were demonstrating their hot air strategy for manned flight, a French academician, Jacques Charles, tested the first hydrogen balloon. The hydrogen (H_2) was produced by reacting zinc with hydrochloric acid ($Zn+2HCl \rightarrow ZnCl_2+H_2$). Though Burke and Wills did pack sixty gallons of rum, carrying hydrochloric acid in glass containers over rough territory was certainly not on. Hydrogen is much harder to contain than warm air, so the silk canopy of *La Charlière* was treated with elastic gum to prevent the gas from escaping. There was also the constant danger of conflagration. Even so, hydrogen balloons were much more versatile than the early 'Montgolfiers'. Jean Pierre Blanchard and John Jeffries floated across the English Channel in 1785, though they lost more hydrogen than expected and had to dump almost everything but an historic airmail packet to avoid ditching in the drink.

Only objects that are lighter than air can float in still air. The molecular mass of air is about seven times greater than that of helium and fourteen times that of hydrogen. Gliders and birds can soar on thermals but, being heavier than air, they cannot simply float off the ground. Anyone who has watched a large bird such as a pelican take off will recognise that it requires a great deal of muscle power and energy expenditure. Fixed-wing gliders must be towed into the air by a powered plane, a winch or an automobile, whereas the much simpler human hang-gliders first climb a hill, then either run down a slope or jump from a cliff.

The lighter-than-air-floating equation is only absolute in the absence of wind. The earth both heats and cools more quickly than the sea, so the winds will tend to blow towards the sea during the day and from the sea at night. Wind currents are much more complex than that, as they reflect global as well as local effects; the air at the equator will always be much warmer than that at the poles. Atmospheric wind contributes to making the east–west flight across the Atlantic about an hour longer than the west–east equivalent. Strong winds can cause a grounded, untethered glider or light plane to lift, then smash back to the earth. Every small boy knows that the captains of aircraft carriers turn their ships into the wind for added lift, relying on both steam catapults and the force of nature to help launch their planes.

Air also carries many small particles. The inland regions crossed by the Burke and Wills expedition are known for their severe dust storms and the explorers would also have seen willy-willys—'mini twisters' that seem to start spontaneously, then move quickly across the plain. The conditions

7

where streams of cold and warm air meet to create these whirling spirals can also cause the much more devastating tornados. Perhaps because of the low population densities on the dry plains of inland Australia, tornados haven't been associated in people's minds with the extreme damage that is all too commonplace in the southern and central United States of America. There are indications that this situation may be changing with global warming. 'Tornado-like' hits have recently been recorded in the coastal suburbs of Sydney and Melbourne.

A tornado can throw a heavy automobile high in the air like a child's toy. Driving along straight roads in the Mississippi delta, you will sometimes see what look to be old-fashioned radio horns mounted on poles by the side of the road. When these sirens sound, the best survival strategy is to leave the car and go to ground in the nearest ditch or bolt-hole. Living in Memphis, Tennessee, Saturday lunch was announced by the midday test sounding of the tornado sirens that are mounted on every fire- or school-house. On a few occasions, we spent part of the early evening in the basement as we waited for the wail of these sirens ('syreens' in Memphis) to cease. A tornado cuts through a neighbourhood like a wide-bladed buzz saw. People die when their insubstantial trailer homes or seemingly magnificent but lightly built houses of framed pine, compressed board and stucco are ripped to shreds in developments located where previous generations spoke of 'tornado alley'. The gods of wind should be treated with great respect.

Many plant species use air movement to spread across the landscape. Breezes blow the seed heads off dandelions, then transport the mini parachutes that contain their genetic material over a wide area. Propeller-like vanes turn some seed types in the wind and greatly increase their range of travel.

The air in hot, humid cities like Memphis and Brisbane is full of pollen and fungal spores that can, for too many, cause annual bouts of asthma. One asthma-inducing component that has become much less common in the air we all breathe is cigarette smoke. In the poorer countries of our planet, the cloud of pollution given off by burning dried cow dung, camel dung, wood or charcoal in poorly ventilated interiors is a major contributor to the early onset of the respiratory condition, chronic obstructive pulmonary disease. Incomplete combustion in inefficient grates and stoves makes these heat sources fifty times more noxious than natural gas. A May 2001 bulletin of the Indian Council for Medical Research states that burning cow dung produces carbon monoxide, suspended particulate matter, the production of polyaromatic hydrocarbons, formaldehyde and a number of other compounds. Apart from direct, toxic effects on delicate respiratory tissue and the accumulation of airborne junk in long-lived lung scavenger cells (macrophages), derivatives of the poly-aromatic hydrocarbons are, like the coal tars in cigarette smoke, known carcinogens—or cancer inducers.

For those fortunate enough to live in the advanced world the switch to electricity generated at remote sites as the principal provider of domestic heat and light has largely eliminated this source of domestic pollution and respiratory damage. The chimney pots on our Victorian house are capped because we no longer use the elegant, coal-burning iron grates that also served as portals to funnel dust and debris into our living space. The natural gas that fuels our current cooking stove and heating furnace is the cleanest burning of all the fossil hydrocarbons. Breathing air so much cleaner than that available to our urban predecessors of even fifty years ago may, perhaps, have made us less aware of the insidious, progressive atmospheric pollution that we now need to address as we seek to combat global warming.

Spores, pollen and seeds can be removed by pumping air through appropriately sized filters, but what of less tangible things? We say colloquially that laughter and music float in air but that's not strictly true. Sound is transmitted through the atmosphere by longitudinal pressure waves that originate in the vibration of human laryngeal cords, or the cone of a loudspeaker, and induce the same vibrations in the membranes of our eardrums. Tiny membrane-attached bones—the hammer, the anvil and the stirrup—pass the perturbation on to the fluid-filled middle ear, where the inner surface of the cochlea is lined with large numbers of hair-like nerve cells. Movement triggers these highly specialised transmitters to send electrical impulses along the nerves. The signals register in the temporal lobe of the brain, where that marvellous central processing unit in our skull somehow converts the incoming information into the perception of noise, words or music.

There is no sound in a vacuum, but is this also true for other invisible entities that many believe are floating in the air around us? Can ghosts and spirits exist in a vacuum? Perhaps we should leave that to Hollywood directors and to the acolytes of belief systems that insist on the existence of the supernatural. Caspar the friendly ghost and the gallant Cavalier carrying his head under his arm can, if the movies are to be believed, float straight through solid walls. The ghosts created by Hollywood special effects artists manifest as transparency and light, but light does not travel through plaster boards, though a high intensity laser beam allows some photons to penetrate through human or mouse skin to the tissues below. When the door is closed, we poor mortals need ducts to get hot or cold air from one room to another.

The ethereal manifestations of the topless Cavalier and Caspar are clearly exempt from the rules of substance, even when the elements of that substance are as minuscule and dispersed as the photons of light or the molecules of oxygen, nitrogen and carbon dioxide in the air around us. If light is a substance, it doesn't float in air, though rays of sunlight illuminate air-borne dust particles. Do the souls of those departed take up airspace? That question is about as useful as asking: 'How many angels can dance on the head of a pin?' The mystics speak of the 'light of the soul', but are they talking about the 'light of the mind'? The rationalist would argue that the spirit world lives only in our perceptions. If we accept that consciousness is a state of mind, that mind is a function of the brain, and that the brain dies in the absence of oxygen, then the spirits are in one sense, like us, 'floating in air'.

As I wrote this, I played mentally with mind pictures of Burke and Wills, Ned Kelly and the Montgolfier brothers. In a sense, I invoked their spirits. All died long before I was born, but they are more real to me than people I meet briefly but barely register because the encounter sets up no associations, no themes in my head. Such memories aren't of the individuals themselves, but of stories, words, paintings or movies. My imagined view of Burke and Wills is conditioned by reading about them, knowing some of the country that they traversed and, not insignificantly, from viewing Sidney Nolan's paintings that conveyed his vision of the two explorers.

Without words and images there are no stories, no spirits floating in the air. As the novelist Tom Wolfe points out, the biblical statement 'In the beginning was the word' can be taken to mean that the birth of the established religions coincides with the development of written language. Does that view of belief make spirituality any less significant or real? Air sustains us, but human beings are not just literal, rational, oxygen-breathing, chemical machines. Some of us are clearly more sensitive to the life of the spirit than others, a predisposition that may even be encoded in our genes.

What lifts the human spirit more than the sudden, unexpected roar that causes us to look up to see the flaming burner and bright mass of a hot air balloon? We smile like children as these harmless, lightweight giants and their human cargos in antique, wicker baskets float gently above our heads. It must have been an incredible experience for the eighteenth-century Parisians who ascended in the first Montgolfier and hydrogen balloons. Those watching could only have been filled with wonder. Were they also a little fearful? After all, unless the story of Daedalus (the father of Icarus) is true, no member of our unfeathered species had ever flown free before and lived to tell the tale. Like the ancients who built the biblical tower of Babel to reach the heavens, would flight invoke the wrath of the deity? Several hundred years earlier, anyone who attempted to fly in a balloon would probably have burned at the stake. In a sense, balloons are a manifestation of the Enlightenment. So far as we know, neither the flights of the human spirit that occurred at that time nor the ascent of these manned gasbags was offensive to the gods.

We have learned to float above the earth since the eighteenth century, then to fly at great speeds and for enormous distances. We penetrated the heavens, found no golden temples or Elysian pastures, and have since behaved like gods of the air. A jet aircraft takes fewer than three hours to retrace Burke and Wills' 1860 route. Only very recently have we fully recognised that the marvels of flight come with a broader cost for the atmosphere. We are not going to turn our backs on the miracle of flight, but we are the stewards of the natural world and it is our responsibility to waste no time in developing cleaner alternatives for air travel. Hot air balloons can burn liquefied natural gas and produce little CO_2 or other pollutants, but jet planes present a much bigger challenge. Perhaps a very different variant of the hydrogen technology that allowed that first flight across the English Channel will provide the answer.

Current thinking is that, like the heart of Australia crossed by Burke and Wills, the Horn of Africa is likely to become even hotter and drier with the continued accumulation of excess CO_2 and other greenhouse gases in the upper atmosphere. On the other hand, East African countries like Kenya may see improved rainfalls. Kenya has been experiencing drought, though the 'long rains' were good in 2006. Paradoxically, the wetter conditions have resulted in the death of many animals that were already in a very weakened condition. Though there may be good years, climate conditions that are extremely variable and unpredictable compromise both the sustainability of natural flora and fauna and the viability of the animal and plant agriculture that supports village communities, farmers and graziers, like the Masai.

Every animal and plant species lives in its own ideal 'climate envelope' determined by ambient temperature, rainfall and the availability of nutrients. How will global warming affect those equations for the plants and animal of the Serengeti? Will the extraordinary variety and number of African animals that we were privileged to view from a modern-day Montgolfier floating over the Masai Mara still be there 100 years from now? Together with poaching, increased mortality from infectious diseases and the progressive deforestation that has resulted from the ever-increasing growth in human numbers, climate change poses a growing threat to African wildlife. Preserving these animals in their natural habitats is a collective responsibility for all of us, not just for the struggling African nations. What a tragedy it will be for the human spirit if the only way that future generations can see African wildlife is to visit a zoo, if that broad landscape of prancing gazelles, roaming lions and wildebeest herds is no more than recorded images on a television or movie screen.

Alphabet Soup

On egg sandwiches and chemistry

Among my earliest and most vivid memories are images of Miss Annie Power's first-grade class-room at Corinda State School. This began my formal education; the idea of kindergarten or pre-school didn't surface till much later in the Queensland public education system. Recalling the occasional screech of chalk misapplied to a blackboard, the scratch of slate pencils on wood-framed slates or the sound of bare feet scraping on an unsealed, dirt-ingrained floor still sets my teeth on edge. It's salutary to recall that we five-year-olds learned our alphabet by inscribing tablets of stone—shades of Moses and his Ten Commandments, which also figured in our schooling—but the slates saved the trees and were environmentally friendly.

Writing on slates meant that memorisation was very important, as the text was wiped off with a wet cloth or sponge at the end of each lesson. This educational approach followed long-established practices. American children were using slates as early as the 1740s, before the 1788 European settlement of Australia, but US primary education had largely moved to 'lead' (carbon) pencils and paper by the first years of the twentieth century. Children entering Queensland public schools as late as the 1960s still had the slate/slate pencil experience: from stone on stone to silicon chips, keyboards and plasma screens in forty years!

Way back in 1946, Miss Power's open-windowed class-room was often permeated with the smell of hot, new leather and hydrogen sulphide. The H_2S came from lovingly made and packed egg sandwiches incubating in brown, calf-hide satchels hanging on the open veranda outside. When the angle was right, these bags were exposed to the full heat of incident sunlight. I expect we developed an early resistance to the consequences of consuming lunches full of rapidly growing bacteria such as *Escherichia coli* and *Salmonella* species which, encountered later by an over-sanitised, western adult, can cause the gastrointestinal distress known by names such as Montezuma's Revenge.

I recall Miss Power as a thin, short, dark-haired, intense, Catholic lady who, it was rumoured, had never married because her fiancé had been killed in World War I. That may well have been true, and it was certainly the case for our tall, redheaded eighth-grade teacher, Miss Thompson (a Methodist, but equally devout). One difference between Miss Thompson and Miss Power was that, while the towering Miss Thompson was forceful in a generally even-tempered and amiable way, I remember Miss Power as being both volatile and angry. When a little kid got his or her numbers wrong, or could not recall 'a, like an apple on a twig', 'b, like a bouncing ball' or 'c, like a cake with a piece cut out', Miss Power would shout 'No, no, child' and rap the unfortunate mite across the knuckles with a wooden ruler. She didn't discriminate on the basis of sex, and the girls were equally traumatised.

From talking to friends who endured the lower registers of the separate Catholic school system at that time, I gather that Miss Power's pedagogic style was pretty much par for the course. That has all changed now, along with the almost complete disappearance throughout the western world of teaching nuns and priests, who would themselves have been subject to a very disciplined, and at times harsh, training. Some conservatives seem to greatly regret the passing of rote learning as a major educational tool, though I expect that even the most traditional would not want to reinstitute physical punishment for very small children. The experience was certainly memorable, though, and we soon knew our numbers and letters.

When I began a new encounter with a somewhat different version of the alphabet at age thirteen, the experience was both less fraught and equally interesting. Now at Indooroopilly High School, we started our formal training in the world of science, with the gifted and personable Stan Brown introducing us to the mysteries of physics. The letters of Dmitri Mendeleev's Periodic Table of the elements became familiar as we learned inorganic chemistry from the tall, deep-voiced, good-looking, athletic Reg Tickle. I can't recall whether any of the girls in my co-ed high school class ended up being chemists, but I expect that he got their attention.

As with all science, the language of chemistry is international, and the experience of learning the subject must be pretty much the same everywhere. Chemistry suits the way my mind works and it came easily to me, as it had to the young Jewish Italian Primo Levi, studying at the University of Turin some twenty years earlier. This was the 1930s Italy of Benito Mussolini. The academically distinguished, classics-oriented Liceo Massimo d'Azeglio secondary school that Levi attended before matriculation was also noted for the anti-fascist stance taken by many of its faculty, no doubt at considerable personal risk.

If our high school teachers were politicised in the 1950s, it was only in the sense of their total commitment to the idea that every child is entitled to a free, first-class, public education from years one to twelve, providing the 'ladder of opportunity' that is essential for the health of all democratic societies. As the United Negro College Fund has it, 'A mind is a terrible thing to waste'. My teachers were outstanding, but the school was otherwise poorly resourced and the newly built facility was absolutely 'no frills' and basic.

We were occasionally reminded that the classroom we occupied for four years was right over the chemistry prac lab, with its all-glass Kipp's apparatus that was used to produce the pungent gas H_2S by reacting liquid sulphuric acid with the solid ferrous sulphide, $FeS + H_2SO_4 \rightarrow H_2S + FeSO_4$. From time to time an English, history or maths lesson was permeated with an intense odour of rotten eggs that floated up between the cracks of the floorboards. Fortunately, the windows could be opened, and the olfactory system adjusts rapidly to bad smells.

By this stage of my account I run the risk that chemical symbols like 'Fe' (iron) and 'S' (sulphur) may be alienating to those who did not have the good fortune to encounter an inspired high school or college chemistry teacher, or just don't have the type of mind that finds science appealing. But I would ask you to bear with me, even if for just a little. Chemistry has its own romance, as many who have read Primo Levi's much-loved *The Periodic Table* will recognise. Levi interfaces his simplified descriptions of chemical assays and the properties of elements with tales of family and boyhood friends who, as heroes of the resistance, were later murdered by Mussolini's fascists, or of redheaded girls who may or may not have given their favours. He describes how, trying to survive as a very young and impecunious private analyst, he used the primitive H_2S-producing Kipp's apparatus to show the presence of arsenic in sugar provided by an old man who suspected a competitor of trying to poison him. A very accessible account takes us through his reasoning processes when, working as an industrial chemist, he strove to solve some practical problems relating to paint quality and manufacture. Such science may not win Nobel Prizes—the fascists and the Nazis together stole too much of Levi's life for him to be able to find that type of employment—but it is what allows our contemporary world to function so effectively in the material sense.

We are drawn in and intrigued by the unfamiliar humanity and relevance to daily life that Levi confers on the world of the working scientist. The dramatic impact is further enhanced by the story of how he survived imprisonment in the Auschwitz death camp. During the first half of the twentieth century,

German industrial and pharmaceutical chemists were very much to the fore. Working in Frankfurt, for example, Paul Ehrlich discovered the organic arsenical Salvarsan (arsphenamine), the first 'magic bullet' for the treatment of syphilis, while Bayer's Gerhard Domagk found that the sulphonamide Prontosil was effective against *Streptococcus*, and he thus ushered in the pre-penicillin era of broad-spectrum anti-bacterial therapy. The consequence was that aspiring chemists, in particular, were required to learn scientific German, a practice that continued until the 1960s for Australian postgraduate students in the sciences.

For Levi, confined as he had been in an Auschwitz work camp for eleven months from February 1944, the fortunate combination of knowing some German and being a trained chemist resulted in assignment to a laboratory in a plant making synthetic rubber. Everything was in short supply in Hitler's rapidly disintegrating Third Reich. With his friend Alberto, Levi stole some small rods of the alloy iron-cerium from a forgotten jar on a dusty lab shelf and painstakingly ground them down, at great personal risk, to make cigarette lighter flints which they traded for bread. Even then, the fact that Levi survived was very much a matter of chance. Marched off just before the Russians liberated Auschwitz on 27 January 1945, Alberto was never heard from again. Only twenty-four of the 640 Jews and fellow resistance fighters who were transported from Italy on the same train of cattle cars as Primo Levi lived to return home. He was one of the very few for whom the cynical sign over the Auschwitz entry gate, '*Arbeit macht frei*' (work brings freedom), ultimately turned out to be true.

From first acquaintance, Levi loved chemistry because it offered the possibility of finding verifiable truths and solutions to real problems. He was bored by philosophy, and it was obvious to him that revealed religion provided no satisfactory explanation for the natural world. But learning chemistry allowed him to think about how matter was constituted, and to design experiments and assays to test his ideas.

All life is, in the end analysis, dependant on chemical reactions. I didn't stay with the inorganic, analytical chemistry that defined Levi's professional career but turned instead to biology and biochemistry. The language of chemistry—whether it's classified as inorganic, organic, physical or bio—is, however, written in the same script, the letters that identify the elements.

More commonly used examples of this alphabetic shorthand are reasonably familiar. O_2 is oxygen, the gas that gives us life. The greenhouse gas carbon dioxide (CO_2) is present at enhanced concentrations in the air we exhale, CO is carbon monoxide, a 'weaker' greenhouse gas and the potentially lethal component in furnace and automobile exhausts. N_2 is nitrogen, the majority component of our atmosphere that, when forced under pressure into the blood, bubbles out again, giving divers 'the bends' if they surface too quickly. Most of us know that H_2O is water. Some who are old enough to recall the earlier days of dentistry and the Marx brothers may remember N_2O, nitrous oxide, or laughing gas. Those who agonise about diet, or are threatened by osteoporosis, are likely to have in mind that Ca and P, calcium and phosphorus, are the major components of our bones.

Descriptors like 'O', 'H', and 'Ca' are easy to translate as the letters begin the word for the element in question. For others, though, it helps to have learned a little Latin. Ferrum gives us Fe for iron; aurum, Au for gold; argentum, Ag for silver; kalium (K) is potassium and natrium (Na) is sodium. *Chloros*, meaning pale green in ancient Greek, leads to chlorine (Cl), and many are aware that NaCl is table salt, the substance we should use sparingly if we want to minimise the risk of developing hypertension.

Those little subscript numbers tell us about amounts. If you take H_2O and apply an electric current, the volume of hydrogen gas that accumulates in a bell jar inverted over the negative terminal will be twice that of the oxygen produced at the positive terminal. In writing out a chemical reaction like

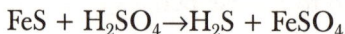

$$FeS + H_2SO_4 \rightarrow H_2S + FeSO_4$$

the numbers need to balance for the different atoms. This is very simplistic chemistry, as it ignores concepts like atomic weight, valency, the contribution made by various charged particles (electrons, protons), the different states of matter and a great deal more.

Even if we remember a few letters that identify key elements (C, H, O, N) and the basic idea that numbers relate to amounts, it makes it much easier to think clearly about the debate concerning the risks of consuming such vast quantities of stored energy by burning fossilised hydrocarbons. It's a fact, though, that while we regard someone who knows little of Shakespeare, word use or sentence structure as uneducated, complete ignorance of a few simple concepts and chemical symbols that define the grammar of life is in no sense considered a disqualification for high office, political power or intellectual prestige. Perhaps that's a condemnation of the way chemistry has traditionally been introduced to those who aren't going to pursue a life in science.

Focusing on areas that fascinate (such as forensics) or on day-to-day examples (how chlorine-based disinfectants work) can help to convey some essential, basic principles in a more generally accessible and interesting way. Alternatively, the negative response to chemistry may reflect the pervasiveness of CP Snow's 'two culture' idea that seems to have convinced many of the more literary that it's impossible to swing both ways (even a little) when it comes to the arts and sciences. That is, I believe, particularly a consequence of the 'British' type of educational system, where fifteen- or sixteen-year-olds can be asked to choose between the humanities and the sciences as their major focus. On the other hand, the continental European model that is reflected in the US ideal of a 'liberal' education encourages young people to maintain a broader spectrum of interests through to, at least, the early college years.

Good, inspired and dedicated teachers are enormously important. Miss Power did the job that was expected of her and did not fail her very young charges when it came to instilling the basics of numeracy and literacy. However, reflecting on a behaviour pattern that indicates a high level of underlying emotional stress, I think she might have been better deployed teaching older children. Many schools had little choice. After the slaughter of World War I and then II, teachers were in very short supply when I started my education. The numbers crisis continues in this most financially under-rewarded, but tremendously important, profession. Particularly when it comes to the 'hard' sciences like physics and chemistry, we continue to be in desperate need of inspired science teachers like Stan Brown and Reg Tickle— and one Professor P, who gained the affection and respect of Primo Levi. Perhaps reading *The Periodic Table* should be an absolute requirement for every academically strong high school and college student.

Life, Gas and Hydrocarbons

At the centre of everything

Energy can be neither created nor destroyed: alive or dead, all plants and animals are subject to this first law of thermodynamics. Life consumes and gives energy. The foods we eat, together with the wood, oil and natural gas we burn for cooking, for heating and to provide transport, are ultimately the products of biological processes driven by the radiant energy of the sun. The capacity for life itself is ultimately derived from the sun that lights and heats our small planet. We recycle the energy that we utilise as we live and then die. Our little march spans that transition from ashes to ashes, dust to dust, with the black element carbon being the principal organiser. Biological systems like the human organism are extraordinarily complex and wonderful carbon-based chemical machines. Carbon takes many forms in nature, including the graphite that goes to make lead pencils and the diamonds that we use to celebrate beauty and commitment, but the most important role for carbon so far as we are concerned is to serve as the scaffold for life. The chemical symbol for carbon is 'C', the centre of everything. The organic compounds that form the basic structures of biology consist of the big C attached to many other molecules like hydrogen (H), oxygen (O) and nitrogen (N).

Though it would be an imposition to inflict any detailed chemistry on the non-scientist reader, making a little use of the shorthand of chemical symbols and the subscripts that denote the numbers of atoms of each type does, as already mentioned, help to give a very broad sense of how different compounds are related. That allows us to simplify the discussion of how, for example, greenhouse gases, like carbon dioxide and methane, are related to living organisms. Skip this symbolic language if you were turned off by high school chemistry and find this to be an annoying distraction. Alternatively, think of yourself as some kind of anthropologist who is trying to gain some glimpse of a different and silent culture by reading the glyphs and runes written in the language C, H, O and N.

When it comes to chemistry, human beings are typical, air-breathing vertebrates. If we ignore the large amounts of calcium (Ca) and phosphorous (P) in our bones we are, in the chemical sense, not too different from any other life form. In fact, we share chemical mechanisms (many of which use Ca and P) that drive cellular functions in organisms as diverse as baker's yeast and fruit flies. The four most abundant chemical elements of our bodies are oxygen (O, 65 per cent), carbon (C, 18 per cent), hydrogen (H, 10 per cent) and nitrogen (N, 3 per cent). Carbon is the basic unit of all organic molecules, water (H_2O) accounts for 55 to 75 per cent of our body mass, and nitrogen is an essential component of the protein building blocks of tissues like muscle, liver and the cells in blood. The variation in water content reflects that women have more water than men, while babies are the wettest of all, which makes a lot of sense to anyone who has ever had to change a squalling infant.

The normal, dry air that we breathe is about 78 per cent nitrogen (N_2), 21 per cent oxygen (O_2), 1 per cent argon (Ar), and somewhere around 0.04 per cent carbon dioxide (CO_2), together with various other minor constituents. Argon is an inert gas that is great for filling light bulbs but plays no role in biology. Otherwise, all the major constituents of the earth's atmosphere are in some way used by the biota (all the life) of our planet.

The haemoglobin protein in the red blood cells (erythrocytes) that circulate through the delicate capillaries of our lungs takes up oxygen for transport to our tissues and organs. The oxygen is then released and consumed within the cells of the body by various oxidative processes that are exothermic (generate energy as heat) and result in the formation of carbon dioxide. Expired air contains about 4.5 per cent CO_2, and the average person breathes out about 450 litres of CO_2 each day. We cannot live without oxygen. We suffocate—or, more correctly, die from anoxia—when, for instance, exposure to excess carbon monoxide (CO) in automobile exhaust fumes both diminishes the oxygen-carrying capacity of haemoglobin and poisons the essential oxidative processes within cells. Both CO and CO_2 are odourless, colourless gases.

Plants are the other half of the O_2/CO_2 equation. Using the energy from sunlight, the process of photosynthesis creates new organic material by combining CO_2 with water to release O_2 back into the atmosphere. For example, the interaction of light with the green pigment chlorophyll in a plant leaf provides the energy to drive the equation $6CO_2 + 6H_2O \rightarrow C_6H_{12}O_6 + 6O_2$, the reverse of what happens in our cells as we generate CO_2. The more familiar name for the simple carbohydrate $C_6H_{12}O_6$ is glucose, while $C_{12}H_{22}O_{11}$ is the sucrose, or cane sugar, that we add to our tea and coffee. Glucose transits readily across the intestinal wall to be carried via the blood to the liver, where it contributes to the assembly of the amino acids (containing N, C and O) that make up the proteins, the fatty acids (fat) and the body's primary energy store, glycogen (both comprised of C, H and O). Blood glucose levels are regulated by insulin and the excess glucose that is associated with obesity can lead to type 2 diabetes.

All the food that we eat depends in some way on various photosynthetic processes. Animals eat plants, ranging from the phytoplankton in the sea to the grasses of pastures and the savannahs. Big animals eat little animals. We humans are omnivores and eat just about everything. In addition, animals and plants can't process N_2 from the air and depend on the capacity of specialised bacteria and phytoplankton (in the sea) to 'fix' nitrogen to form the ammonium (NH_4) compounds that we use to make proteins. Legumes, like peanuts, mung beans, lentils and sweetpeas, have nodules containing the bacterium *Rhizobium* that does the job of fixing nitrogen. The legume alfalfa is often ploughed back into fields to provide nitrogen for other plants. Otherwise, plants need nitrogenous fertilisers. At the other end of the cycle, denitrifying bacteria break down rotting animal and plant matter to return N_2 to the atmosphere.

Plants and plankton provided the majority of the starting material for the vast reserves of fossil fuels that we've been consuming with such abandon over the past 300 years. Most of this enormous energy store originated in the small ocean organisms and the trees and ferns of the cretaceous period that occurred something like 360 million years ago. The bodies of dead animals also made some minor contribution. These ancient sea-beds, forests and marshlands were, as a result of volcanic activity and the shifting of tectonic plates, compacted under layers of earth and rock, where they were then exposed to the action of anaerobic bacteria—bacteria that function in the absence of oxygen. The result was the formation of the fossilised hydrocarbons: coal, oil and natural gas.

We burn these fossil fuels to generate energy as heat. In chemical terms, the positively charged oxygen takes electrons from the negatively charged hydrogen in an exothermic reaction that continues till either the fuel source is exhausted (transformed into the oxidised form) or the oxygen is consumed. Some form of ignition, ranging from the flame of a phosphorous match or pilot light to the electric discharge of a lightning strike, is generally required to start the process. Forensic firemen and insurance inspectors look for highly combustible 'accelerants', like petroleum, in suspected cases of arson.

Both coal and oil are complex mixtures of C and H, while the 'fossil fuel' gases are defined compounds. Coal is commonly mixed with sand and clay, with the cleanest burning form being the shiny, black anthracite that has the highest content of pure carbon. Oil, which is processed in various ways to give diesel, petroleum and so forth, came into general use only with the 1859 distillation of kerosene (a complex mixture of C_9 and C_{14} hydrocarbons) by the Canadian geologist Abraham Gesner. Natural gas is mostly methane (CH_4) but it also contains propane (C_3H_8), butane (C_4H_{10}) and ethane (C_2H_6). Methane, or marsh gas, is a prominent greenhouse gas in its own right. It is also produced in those large 'fermentation vats', the fore-stomachs of ruminants such as cattle and sheep.

With one carbon to four hydrogen atoms, methane burns with the production of relatively little carbon dioxide. Buses in Washington DC announce prominently that they are fuelled by clean-burning natural gas. Hong Kong, with its high levels of atmospheric pollution, is also moving in this direction. Methane-powered automobiles were a not uncommon sight in World War II Australia—although, at that stage, the country was unaware that it was sitting next to very large reserves of natural gas. Inventive owners fitted normal cars with homemade 'gas producers', often a large cylinder containing (for example) fermenting chicken manure. The resultant methane accumulated slowly in a huge, rubber bladder on the car's roof. Neither the performance nor the range of these Heath-Robinson vehicles was exactly stellar.

Methane produced from the fermentation of biological garbage is a potential energy resource that is currently grossly under-utilised. As the greenhouse effect of methane is about twenty-three times worse than CO_2, it makes sense for us to burn this methane to generate heat as energy. The Sierra Club is active in advocating the expansion of such programs, particularly for application on farms that generate a lot of animal waste. A BMW plant in Spartanburg, South Carolina, aims to power a new paint shop using piped methane derived from the natural fermentation of garbage in a nearby landfill. The state government of South Carolina is being pro-active in encouraging the wider use of methane for electricity generation.

By 2003, more than 26 per cent of the methane produced in large Australian landfills was being recovered and used largely for electricity generation. The Western (sewage) Treatment Plant at Werribee near Melbourne uses membrane-covered fermentation lagoons to trap methane that is then burned to boil water and drive a steam turbine. Some analysts are speculating that the small-scale, domestic-use principle exemplified by the automobile gas producers of the 1940s could help solve the problem of how to deal with the 10 million-odd tonnes of dog and cat waste that is generated in the United States every year. Maybe those gas producers weren't so great on cars, but could they contribute to household heating?

When it comes to motor fuels derived directly from recently generated (rather than fossilised) hydrocarbons, much of the current focus is on the liquid form of ethane, ethanol (C_2H_6O). The ethanol produced by bacterial fermentation of plant products in whisky distilleries and the like is what traffic police are looking for when we blow into the breathalyser; although breathalysers will, in fact, detect any volatile compound carrying a methyl (CH_3) group. Methanol (CH_3OH), or wood alcohol, can also fuel engines ($2CH_3OH + 3O_2 \rightarrow 2CO_2 + 4H_2O$), but it is much more toxic than ethanol, and street alcoholics who (knowingly or otherwise) consume methanol are in danger of going blind, or worse.

Large-scale industrial manufacture of ethanol from sugar cane, corn or any other plant material that may come to hand is increasingly focused on powering motor vehicles rather than on providing their owners with supplies of rum, whisky, bourbon and moonshine. Currently, manufactured cars with modified 'flexible fuel' engines will run on up to 85 per cent ethanol. Although burning ethanol in the cylinders of an internal combustion engine still contributes to the greenhouse effect ($C_2H_6O + 3O_2 \rightarrow 2CO_2 + 3H_2O$), it does so to a much lesser extent than when we use petroleum. With its 1:3 carbon to hydrogen ratio, ethanol (C_2H_6O) is almost as good at the 1:4 of methane (CH_4) when it comes to limiting greenhouse gas emissions.

The extensive plantings that are required for mass ethanol production function also as 'photosynthesis factories' to take CO_2 out of the air. Where tensions can emerge is between the need for cheap food, particularly in developing countries, and the desire to use biomass for fuel production. However, this could cut both ways. Mexican farmers are currently suffering from the 'dumping' of cheap US corn in their market, which is permitted under the North American Free Trade Agreement (NAFTA). If much of that corn were to go towards ethanol production, it could help the economic viability of local growers.

Diesel engines can be tuned to run on almost any vegetable oil. One Memphis company, Deep Fried Rides, for instance, cleans up used cooking oil by removing the fat and other impurities (bits of catfish), then sells it for consumption by diesel-powered automobiles that have been converted at a cost of a couple of thousand dollars or so. If I had a diesel car and didn't try to walk everywhere, I could consider joining the Melbourne Biodiesel Club. There are concerns that the widespread use of plant products would decrease the production of food for human use but, as farmers are being paid not to plant in the United States and the sugar cane industry everywhere is looking for new markets, there is certainly room for substantial expansion.

To return to the fossilised hydrocarbons, serious coal mining only began within the past 500 years and, though there is evidence that the Chinese were drilling for oil 1500 years ago, the first significant North American oil wells date back to the 1850s when it was realised that kerosene was an effective substitute for whale oil in lamps. Before that, wood and charcoal were the primary heat sources for cooking, household heating and the forges used by blacksmiths. Charcoal is made by 'carbonising' wood. Conical heaps of cut timber are covered with turf and soil, leaving an opening at the bottom and the top that provides sufficient air flow to allow some of the wood in the centre of the pile to burn. The resultant heat drives off the water and volatile substances from the remainder of the wood that fails to combust because of the lack of available oxygen. The mass of the wood is thus reduced by about 40 per cent to provide a high-quality fuel. Charcoal burning is a skilled craft that is still practised in some regions of Africa, with much of the product being used in the Arab countries to the north. Unfortunately, this traditional practice contributes to deforestation and the soil erosion and degradation of the landscape that follows.

Burning wood, charcoal, coal, oil or natural gas in some form of stove or furnace is inevitably accompanied by the production of both carbon dioxide and carbon monoxide, requiring a chimney or some other form of flue to vent these gases and other volatile products out of domestic living spaces. The great eighteenth-century American diplomat and scientist Ben Franklin understood political realities and lightning, but his first stand-alone iron stove had a flue system that was too complex and did not draw very well. The early versions tended to fill the room with smoke and the stove soon went out. This fault was rapidly corrected by another famous Philadelphia identity, David Rittenhouse. Rittenhouse has an elegant square named after him, but Franklin gets the credit for the highly successful, contemporary versions of both his stove and his lightning conductor.

The energy released by the combustion of various hydrocarbons has thus been a major determinant of progress and wellbeing over the centuries. For thousands of years, burning wood, and then charcoal, has allowed us to cook, heat our living spaces and forge the metals to make household utensils, agricultural implements and the weapons and armaments of war. The transformation in the human condition that began less than 300 years ago with the Industrial Revolution was, however, only possible because we learned how to access the enormous, stored energy potential of subterranean fossil fuels.

The plain fact of the matter is this: fossilised hydrocarbons are irreplaceable resources. The sooner we stop burning the remaining reserves, the better off we will ultimately be. Oil is also used for a variety of other purposes, including the manufacture of plastics. We are consuming the future, and it's up to us to develop and use renewable alternatives. So what if there is an 800-year reserve of coal? Human beings have been around for at least 100 000 years, and most of us want to see that continue for much longer than another millennium or two.

Soaring with Eagles

On turkeys and Aryan demagogues

Of all the renowned scientific institutes of the world, Louis Kahn's architecturally brilliant Salk Institute is the only one that invariably causes my spirits to soar, even before I walk into the building and have those intriguing conversations about new and exciting research results and ideas. Scientific research institutes tend to be utilitarian buildings that, at best, look like stylish, modern office blocks and, at worst, resemble mid-twentieth-century industrial sites or hospitals. The Salk Institute, named for the great Jonas of polio virus vaccine fame, is an exception. Sited high on cliffs in La Jolla, California, the tower offices, where communicators like Joseph Bronowski and Nobel scientists Francis Crick, Roger Guillemin, Sydney Brenner and Renato Dulbecco wrote and thought, look west to the blue Pacific. The open spaces are organised so that the buildings blend naturally to give an elegant, unobtrusive aspect.

At the Salk not so long back, I took time out to stroll across to the adjacent headland and watch the hang-gliders flying high above Black's Beach. They swoop, then soar as the dangling pilots catch thermals, those columns of warm air that carry seagulls and pelicans up into space. I haven't tried hang-gliding but, like many recreational activities that involve an element of real risk to life, it's easy to see that the spirits of the rapidly lifting flyer must also soar. There's a bad joke (untrue) that the Salk Institute offers free hang-gliding lessons to senior scientists who aren't too self-aware but have reached the age at which, having lived by their wits, it's inevitably time to move on and out. Down below, on Black's Beach, the nude bathers can, at least on a bright summer's day, have another sort of full exposure to hot air that presumably gives them some sort of a lift.

On a summer evening in southern Scotland, we sat and watched as fixed-wing gliders soared above the green and gentle landscape. My Edinburgh boss, Dick Barlow, who taught me both the intricacies of neuropathology and how to write clear, concise scientific English, was a glider pilot. So is Paul Nurse, the current president of the Rockefeller University. Paul is an adventurous soul who also flies powered aircraft. We lost a very good friend and colleague, the Hamburg virologist Fritz Lehmann Grube, when, not long after he retired as director of the Heinrich Pette Institute, his glider crashed into the side of the Alps.

After hot-air balloons the glider was the first machine that allowed humans to launch into thin air and survive. Leonardo da Vinci designed a glider, but he would probably have been in danger of the Holy Office of the Inquisition and the fire if he'd tried to test it in mediaeval times. The early gliders of the German Otto Lillienthal were both too primitive to soar with thermals and very dangerous. Lillienthal died in 1896, after he fell from a height of seventeen metres and broke his spine. A true scientist, he evidently said '*Opfer müssen gebracht werden*' ('Sacrifices must be made'), which may be marginally acceptable, though much to be regretted, so long as the person to be sacrificed is the experimenter. Though I haven't visited the National Soaring Museum on Harris Hill at Elmira, New York, evidently it traces the development of the elegant gliders with their broad, drooping wingspans that we know today. From 1930 to 1946, Harris Hill was the site of the first National Soaring Contests.

When we think of soaring birds, our minds turn naturally to that most powerful of all avian predators, the eagle. While living in Tennessee, we made an occasional winter excursion to Reelfoot Lake, the protected home of some two hundred American bald eagles. When the tectonic plates that meet at the New Madrid fault line running west of Memphis last moved substantially, in 1812, the disruption was so great that the mighty Mississippi River flowed backwards for a time to form Reelfoot Lake. Very few people lived in western Tennessee at that time but, when the New Madrid fault shifts again, the consequences could be disastrous, with massive damage to cities and towns. As a consequence, the Center for Earthquake Research and Information at the University of Memphis has a constant monitoring brief that is taken very seriously indeed by all who live in the region.

The Australian federal capital, Canberra, has a tall column bearing a proud, grey eagle that was erected by a grateful nation to commemorate American sacrifices in the Pacific during World War II. The eagle did not, however, become the icon of the thirteen American colonies that declared their independence from Britain in 1776 without some debate. Among the founders of the Republic, Philadelphia's Ben Franklin was very sceptical. Writing in the style of another age, he argued that the turkey would be a much more appropriate symbol:

> For my own part I wish the Bald Eagle had not been chosen the Representative of our Country. He is a Bird of bad moral Character. He does not get his Living honestly. You may have seen him perched on some dead Tree near the River, where, too lazy to fish for himself, he watches the Labour of the Fishing Hawk; and when that diligent Bird has at length taken a Fish, and is bearing it to his Nest for the Support of his Mate and young Ones, the Bald Eagle pursues him and takes it from him ... For the Truth the Turkey is in Comparison a much more respectable Bird, and withal a true original Native of America ... He is besides, though a little vain & silly, a Bird of Courage, and would not hesitate to attack a Grenadier of the British Guards who should presume to invade his Farm Yard with a red Coat on.

Apart from being an eagle critic, Franklin is famous for having invented the lightning conductor that defuses bolts of electricity by directing them safely to earth so that they do not start fires. The imperial eagle, or Aquila, topped the standard carried at the head of the Roman legions. The bronze eagle with outstretched wings arches above stylised lightning bolts, the fire of the gods, and the bold letters SPQR: *Senatus Populesque Romanus*, the Senate and the Roman people. The 'people' were the Roman upper classes. We still use the term 'aquiline', like an eagle, to describe the classic aristocratic visage.

The heraldic device of the Byzantine Empire, then Charlemagne's Holy Roman Empire, was a double-headed, triple-crowned eagle, with a sword or sceptre in one hand and an orb in the other. This symbolised that the inheritors of Rome's power looked both east and west and it was later assumed by the Hapsburgs in the declining days of what had by then become the Austro–Hungarian Empire. The double-headed eagle was also used in the Crusades, by the Freemasons and by Ivan III of Russia. Not too biologically accurate, but looking impressive on a flag, the all-seeing all-powerful eagle soars above, ready to swoop with lightning speed when errant peasants, intellectuals or even aristocrats get out of line. By having two heads, the eagle seemed to have achieved sufficient power to dispense with the accompanying lightning bolts.

The vaunting symbolism of a stylised eagle, now mounted above a swastika, was co-opted by the short-lived thousand-year Reich. The lightning bolts of the Roman Aquila transformed into the double 'Sig Rune' SS insignia of Heinrich Himmler's black-shirted praetorian guard. In the Norse mythology that the Nazis co-opted for their own ideological purposes, the runes were symbols carved for magical use by the god Odin, Thor the Thunderer's father.

Adolf Hitler's retreat, the Berghof at Berchtesgaden, Obersalzburg, had an associated 'tea house', the

Kehlsteinhaus (Eagle's Nest), which was meant for entertaining visiting dignitaries. Evidently the great dictator rarely took the 124-metre elevator ride up to the Kehlsteinhaus because he feared heights. Some of Eva Braun's home movies show Hitler as a well-to-do, affable, middle-class man, playing with his dogs, strolling down pleasant paths and joking with friends as he relaxed in his mountain vacation home. *Mein Kampf*, his personal statement written before he came to power, reveals Hitler's simplistic thinking and deep sense of grievance but gives no hint of either a lightning wit or a soaring intellect. As many have remarked, what was so horrifying about the Nazis was their banality, their superficial ordinariness. In some sense, they are so like us.

Though Hitler and Himmler were to wield enormous power that enabled them to devastate nations and murder millions of their fellow human beings, neither was in the least bit aquiline in aspect. Both about 1.75 metres tall, Hitler sported a toothbrush moustache that was very effectively lampooned by Charlie Chaplin, and Himmler had a face that was far from attractive. As is often the case with political demagogues, their ascendance in the popular imagination was all in the mind! Their fellow gang leader, Benito Mussolini, was also no oil painting. On the opposing side, the aristocrat Winston Churchill looked like anybody's baby, while Franklin D Roosevelt flashed a winning smile and Charles de Gaulle had an imposing physical presence. Their ally, Joe Stalin was, it turns out, as monstrous as Adolf, Heinrich and Benito combined, but he did have a better moustache.

Events proved that Hitler was no ordinary human being. Drawing on his own emotional intensity, he tapped into the dissatisfaction and despair caused by the economic collapse in Germany. His rhetorical skills, his capacity to generate inspired hot air, caused the spirits of his followers to soar like eagles catching thermals. Nobody who has seen the young cinematographer Leni Riefenstahl's propaganda masterpiece *The Triumph of the Will* can forget the close-ups of those faces shining with self-confidence and pure delight as their beloved Führer, standing above them on the still existing dais at Nuremberg, draws a vivid word picture of the manifest Aryan destiny open to true believers and the genetically pure. Even those who don't speak German can grasp the power of his oratory. Starting calmly, the rhetoric increases in power, the right arm saws back and forth at body level as if it is trying to take off, and is then released and raised to gesticulate extravagantly as the ranting voice achieves a pitch of harsh intensity. The performance is mesmerising and, because of who it is and what he accomplished, profoundly terrifying rather than ridiculous or even absurd.

The imagery and the tone of the commentaries in the German newsreels made during the Blitzkreig, the lightning war, as the Stukas dived and bombed and the Panzers rolled through Belgium, then France, give a sense of the soaring spirits, the arrogance that pervaded the Über Deutschland of that era. The first setback came when the fighters and bombers of the Luftwaffe were defeated in the Battle of Britain. Most of the young Messerschmitt, Heinkel and Dornier pilots who killed and were killed in the skies over England learned to soar in gliders before they flew powered aircraft. At least in the early years of Nazism, the mandate of the Versailles Treaty that there should be no German air force was still observed. Once their engines are blasted to pieces, military planes are too heavy to float in air and they head only towards the ground. Gasoline, hot engines, incendiary bullets and oxygen make a bad cocktail.

The price that ordinary Germans paid for their surrender to systemic lies and rhetorical hot air, that brief period of soaring confidence and collective cruelty, became increasingly bitter. Following the experience of cities like London and Coventry, the British bombers transformed Dresden into a furnace that consumed almost every living thing. Kurt Vonnegut Jr, the American science fiction writer, lived through the fire-bombing of Dresden in an underground slaughterhouse used to confine prisoners of war. He relates his experience in *Slaughterhouse Five*, a novel that was turned into a perceptive and intriguing movie of the same name. Fire became a sub-text for Vonnegut. He later wrote sympathetically about those great human beings the volunteer firefighters, who are so characteristic of the best in the cultures of rural America and Australia.

Unlike Vonnegut, Adolf Hitler and Eva Braun did not survive World War II. The story of their final hours is told with a calm effectiveness in Bruno Ganz's *Downfall*. After the imperial couple suicided, their bodies were doused with petrol and burned in an open space adjacent to the Berlin bunker that saw the dying days of the Nazi empire. As the recently wed 56-year-old Adolf and 43-year-old Eva were consumed by fire, they joined the millions of men, women, children, babies, grandmothers, grandfathers, husbands, wives, priests, pastors, professors, poets, tailors, gypsies, rabbis, merchants, idiots, bootmakers, intellectuals, writers, illiterates and just about anyone you can think of who were incinerated in bombing raids, or were gassed then vaporised in the crematoria of Auschwitz–Birkenau and like places of horror.

Those who doubt the power of soaring rhetoric, the social conflagration that can result from the corrupting mix of fear, ignorance, racism, intellectual laziness, targeted paranoia and political hot air, should pause and reflect. Think about it next time sensibilities are jolted by one of those 'shock jock' sociopaths who infiltrate the world of talk-back radio. Whether the particular zealot is driven intellectually from the 'right' or the 'left' of the political spectrum, the essence of totalitarianism is that nuanced, balanced, evidence-based discourse is considered 'weak' and has no sway. Like most predators, eagles are not subtle, and they are basically about death.

Some of us are privileged to write editorials or to be public commentators. Anyone can send a letter to a newspaper's editor or ring in to a radio program. But what if you were living seventy years ago and the broadcast network was controlled by Hitler's information minister Joseph Goebbels, or the newspaper was the Nazi *Volkischer Beobachter*, the fascist *Il Popolo d'Italia* or the *Pravda* of Stalin's USSR? If your contemporary text would fit comfortably in any of those formats, then you might just want to think a little, take a good hard look at who you are, what drives you and what you're doing with your life. The Nazis were supported by the mass of 'good' German people. They came to power via a democratic voting process and are not unique in history. The mixture of economic collapse, and the combination of a deluded electorate with business, religious, political and military leaders who are prepared to cede leadership to inspired

44

liars and psychopaths in the interest of 'stability' is one that we have seen before—and will see again.

Berlin has long been one of my favourite cities. As a guest at a rather fancy occasion years ago, I stayed at the very upmarket Kempinski Adlon. The Adlon was Berlin's leading hotel in the first part of the twentieth century but, like most of the central part of the city, was destroyed during the Allied bombing. The reconstruction is very faithful: the woodwork is extraordinary and, as you ascend in the elevator, the floors are identified by an antique hand moving around a clock face rather than by the familiar panel of lights. The surrounding, re-built cityscape is similarly splendid and from that 1920s era.

Strolling around the streets near the Adlon I suddenly saw, for the first time it seemed, the magnificent city that existed when the Nazis came to power. The realisation hit me like a thrown brick: by playing on fear, prejudice and envy, a bunch of sick, physically unattractive but intelligent and well-organised criminals was able to occupy the predecessors of these fine buildings and associated institutions, dismantle German democracy, systematically destroy the liberal, open, well-educated, pluralistic society that characterised 1920s Berlin and trigger a global conflagration. It took my breath away and, because what the Nazis started so damaged both my generation and the lives of our parents and grandparents, I came close to weeping. How easy it is to underestimate the fragility of our governance structures and the power and pervasiveness of evil. How readily we suspend judgement and fall prey to crude, simplistic models that diminish personal freedom and self-expression while imposing draconian solutions for maintaining social stability that hide behind the rationalisation of ensuring public safety. There are clear lessons for our time.

If visiting Berlin doesn't give you a sense of that disastrous period of human history, then take a forty-minute train ride and visit the memorial on the site of the Sachsenhausen concentration camp. I haven't been there, but we spent a day at Auschwitz–Birkenau near another beautiful city, Cracow in Poland. That was enough for one lifetime. If you pick up a pinch of dirt from the roadside, there are little fragments of burnt, white, human bone that vented in the smoke and flame from the tall chimneys of the now destroyed crematoria, then soared for a little in the surrounding air before falling to permanent repose in the earth.

The Nazi era provides a warning that we should never forget. Ben Franklin was right: when it comes to politics, it's much better for nations to be grounded like turkeys than to soar like eagles. We can't avoid the lightning strikes that cause bush fires, but we can take care that we are not seduced by lies, rhetoric and hubristic symbolism into feeding conflagrations that have the potential to burn back and do us great harm.

When we eventually get around to forming an Australian republic, we could benefit from the mistake made by the American secessionists and seize on Ben Franklin's advice that the turkey is a more apt representation of a modest, decent nation than some soaring, avian predator.

Perhaps the Australian brush turkey is an appropriate contender: the male is a great home builder, but it can be a little annoying if this protected bird builds his four-metre wide by one-metre high nest in your back garden. Though, unlike the penguin beloved of social conservatives, he is far from monogamous; the male brush turkey accepts the duty of constructing and maintaining a warm environment to hatch the eggs laid by his female consort(s). Apart from this selective adherence to 'family values', survivors from both the right and left of Australian politics might also have difficulty with the brush turkey's Latin name, *Alectura lathami*. Maybe we should stick with the emu (*Dromus novohollandiae*) that confronts the kangaroo on the current Australian coat of arms. Emus are considered to be about 70 per cent monogamous and, like brush turkeys and penguins, the males incubate the eggs. But emus are flightless, grounded stay-at-homes, and Australians are great travellers, with about 25 per cent of the population having dual citizenship and about 5 per cent being out of the country at any one time.

Then there's the mutton bird, the short-tailed shearwater (*Puffinus tenuirostris*). A prodigious swimmer, this most abundant of all Australia's soaring seabirds undertakes a massive migration between the southern and northern hemispheres each year. As in the late Peter Allen's emotive song, our great avian athlete 'still calls Australia home'. It nests in Tasmania, on the islands of Bass Strait and in other south-eastern locations. Along with the eucalypt, the whale, the elephant seal, the penguin, various snakes and the goanna (lizards, *Varantus* species), boiled-down mutton birds contributed to our earliest entry into the oil industry. They are altogether a graceful, resilient and useful bird. At least to me, the short-tailed shearwater passes Franklin's test for an appropriate national symbol.

Burnt by the Sun

UV

Focusing the midday sun with a magnifying glass causes paper to smoke, then burn. Bushfires start when the sun's rays are concentrated on dry grass by a fragment of broken bottle. Sunlight reflected to a point by a concave metal or glass mirror can boil water or cook meat. The sun dehydrates and burns us if we are exposed for too long without protection (adequate skin pigmentation, a chemical sun block or suitable clothing).

Dangerous and damaging though it can be, we love the sun: some worship it. We sing, 'you are my sunshine, my only sunshine'. The bright gold of aristocratic heraldry is 'or', for the Latin *aurum*; or *sol*, or *soleil*, meaning sun. Our breakfast eggs come sunny-side up, we buy cars with sunroofs and sit on sun decks. Some of us have sunny dispositions. Open to the sun, we recognise its enormous power. Recently though, we have started to become very concerned about global warming, about the possibility that we have, in our self-absorption and arrogance, not being according the sun its due respect, that there will be a price to pay.

Isolated in space on the third planet from the sun we wonder, as humans must always have done, what the future holds for us. We know that the particular mix of water, atmosphere, soils and ambient temperatures that we find on this small world provides the only set of conditions capable of supporting complex life forms in our solar system. There may be other such places elsewhere in the universe, but even if we find out that they exist, how could we ever reach them?

What is this golden orb that shines above by day, then, by the rotation of the earth, hides its face by night? Some four hundred plus years of research in physics and astronomy, going back to people like the Dane Tycho Brahe and his student Johannes Kepler who first proposed the laws of planetary motion, began the process that led to the current perception that our sun is an average-sized star at the centre of a small solar system on the outer edge of the spiral arm of the Milky Way. About 4.5 billion years old, it should remain in its present form for another four to five billion years. That's the sort of time frame we need to have in mind when we talk about human continuity, not the minuscule less than a thousand years or so that we expect coal or uranium supplies to last. We've been very good at digging stuff up and burning it, but now we need to prove that our intellects make us 'smarter' in the evolutionary sense than those long-term, cold-blooded survivors, the cockroaches and the crocodiles.

The sun—a ball of incandescent energy that sits more than 146 million kilometres from us—is a gaseous mass of 74 per cent hydrogen and 25 per cent helium that has a surface temperature of about six thousand degrees Celsius. The pressure of gravity at the sun's core combines hydrogen nuclei to form helium-4 plus excess energy: a fusion reaction that proceeds via 'heavy hydrogen' (deuterium) and helium-3 intermediates and involves positron, proton, neutrino and neutron exchanges. Nuclear physicists can reproduce such processes but the Holy Grail is to make sufficient quantities to serve as a source of perpetual, renewable and relatively clean power. The problem of containment is enormously difficult, but progress is being made in this very dynamic area of nuclear fusion research.

Sunlight reaches us in about 8.3 minutes. The idea that light consists of photons (particles) with frequency-dependent energy characteristics was first proposed in 1905 by Albert Einstein, building on Max Planck's 1900 suggestion that the energy of electromagnetic waves is organised as small bundles or quanta. Einstein is remembered principally for the Theory of Relativity ($e=mc^2$) and his unruly mop of hair. No scientist has ever come closer than that to true immortality in the public mind. Just as familiar to scientists is the Max Planck organisation that run most of the top German research institutes.

Long before Planck and Einstein, Isaac Newton (1642–1727) of falling apple (gravity) fame investigated the property of prisms to split light into the familiar red, yellow, orange, green, indigo and violet. The energy spectrum extends from the longest infrared waves through the visible range to the ultraviolet. The heat that warms and sustains us is mostly derived from the long, infrared end. After filtration through the atmosphere, the total sunlight package exerts about 1000 watts of energy per square metre of the earth's surface. Think of the heat from a 100-watt electric bulb.

This energy sustains us. Without the warmth we get from the sun, our world would be an uninhabitable, frozen rock. Because the sea takes longer than the land to both gain and lose heat, the solar energy differentials that we experience as day and night drive the breezes that cool us and the windmills that, over the past millennium or more, have pumped our water, have ground our grain and are now providing a modicum of electricity to serve our enormous demand for energy. All life depends ultimately on sunlight-induced photosynthesis in plants, a process that has also supplied the fossil fuels that we continue to burn with such reckless abandon.

We are beginning to harness some of the sun's energy directly using relatively unsophisticated technologies, such as solar panels and small desalination plants. A simple form of the latter uses hessian dipped in salt water as an evaporator, then a condenser to recover clean drinking water. The principle is not too different from that of the Coolgardie safe familiar to many older Australians. Electricity generated by, for instance, the capacity of photons to excite electron transfer in crystalline silicon semiconductors is already finding widespread small and large-scale application. Solar-powered watches and battery lamps are readily available. With well-resourced research and development efforts, perhaps funded by carbon taxes on the sale and subsequent combustion of fossil fuels, we can expect to see the emergence of many novel approaches that eliminate the hydrocarbon 'middle-man' by enabling the immediate harvesting of solar energy for human use.

Apart from the fact that supplies are finite and will eventually become exhausted, many of us now acknowledge the problem that we face as we continue to burn CO_2-emitting, fossilised hydrocarbons. If the process of radiating some of the sun's heat energy back into space is inhibited by the accumulation of gases like CO_2 and methane (CH_4) in the atmosphere, then we get the 'greenhouse' effect that results in global warming. Greenhouses are great, so long as you can control the temperature range, but the sun doesn't have a thermostat that we can access. Being in a sauna or hot tub that's turned up too high is an experience that soon becomes both unpleasant and tiring. It can even be lethal if, as in the occasional murder mystery plot, we are locked in and can't escape—the situation for all of us on sauna-earth.

James Lovelock, the British originator of the Gaia hypothesis, suggests that we could buy some time to cool the planet by making a form of atmospheric sunshade. The proposal is that adding sulphur (S) to jet fuel will convert the 'contrails'—the clouds of ice crystals that form around particles in the exhaust of high-flying jets—to droplets of sulphuric acid (H_2SO_4). Though sulphur is currently removed from jet fuel, North America evidently warmed a little when all the planes were grounded in the immediate aftermath of the September 11 attack in 2001. The acid sunscreen idea is being discussed by those who think about atmospheric physics, although, if my personal reaction is typical, the approach sounds a bit terrifying to anyone who ever had to handle one of those intimidating, big brown bottles of H_2SO_4 in a high school chemistry lab. Add the fuming acid to the water, not the other way round, right?

The main concern about Lovelock's idea is that, while minute 'acid drops' might function high in the stratosphere to reflect a modicum of incoming solar radiation back into space, gravity's pull would eventually bring at least some down to the earth's surface as acid rain. The earth has experienced global sunscreens before and will probably do so again. A hit by a large asteroid that threw up a massive dust cloud and caused dramatic global cooling is one of the explanations raised for the sudden, mass extinction of (at least) the large dinosaurs some 65 million years ago. The case that the descendants of some small dinosaurs are still with us as birds looks increasingly convincing. The ancestral crocodiles and cockroaches survived, as did the rodent-sized mammals that had evolved by that time. Unlike crocodiles and cockroaches, our big-brained species *Homo sapiens* has only been around for a maximum of about 200 000 years.

Long before we understood the nature of solar radiation, the heat and light of the sun's bright rays figured prominently in our view of life, love and nature, and our place in the world. Sunlight illuminates the images captured by painters like Sidney Nolan or JMW Turner—harshly in the case of Nolan, who was dealing with the burned-off Australian landscape; and gently in Turner's Venetian and English canal and seascapes. The sun is a recurrent theme in the works of great poets. Shakespeare's eighteenth sonnet will be familiar to many:

Rough winds do shake the darling buds of May,
And summer's lease has all too short a date:
Sometimes too hot the eye of heaven shines,
And often is his gold complexion dimmed …

WH Auden, who also grew up in that gentle English land-scape, had the relationship between plant life and the sun right when he wrote in 'Law Like Love':

Law, say the gardeners, is the sun,
Law is the one
All gardeners obey
To-morrow, yesterday, to-day.

A more threatening and ominous presence links the dramatic narrative in Coleridge's *Rime of the Ancient Mariner*. Among many examples:

All in a hot and copper sky,
The bloody Sun, at noon,
Right up above the mast did stand,
No bigger than the Moon.

The 'hot and copper sky' and 'no bigger than the moon' suggest that the sunlight shining on the Mariner's ship was partially blocked by dust haze. Perhaps they were sailing close to one of the desert coasts. What he then writes could be taken as a prediction of three likely consequences of global warming, rising ocean levels, damage to man-made structures and periodic drought:

Water, water, every where,
And all the boards did shrink;
Water, water, every where,
Nor any drop to drink.

We may already be experiencing that latter effect in parts of Africa and Australia. The relationship between the burning heat of the sun and the unpredictability of rainfall is a constant Australian theme. Every kid of my generation learned Dorothea McKellar's iconic 'My Country':

I love a sunburnt country, a land of sweeping plains,
Of rugged mountain ranges, of droughts and flooding
    rains
[...]
Core of my heart, my country! Her pitiless blue sky,
When, sick at heart, around us we see the cattle die ...

The heat and dryness that once had the greatest impact on rural agricultural communities has—with increasing population, decreasing rainfall, water rationing and the threat of bushfires in the outer, leafy suburbs of the big cities—become a menacing reality for all Australians, as well as for many Americans in states like California and Colorado.

Recognition of the sun's power contributed to social and religious practices that date back to ancient times. Sun worship continues, but the priests and acolytes of the modern Sun God are very different from those authority figures that sourced their power from the imagined Ra, Horus, Helios, Apollo, Ravi or Apu Inti. Instead of practising rituals that might involve long robes, sonorous pronouncements, murderous practices with stone knives and obscurantist mumbo jumbo associated with star gazing and examining the entrails of unfortunate animals, today's sun worshipers fly in jumbo jets, wear shorts and bikinis, tote surfboards and don't feel the need to dominate anyone.

Some build a whole spiritual ethos around sun, surf and sand, but (as with any belief system) the majority are less dedicated and mix their desire to experience warm water and balmy days and nights with activities like the consumption of colourful mixtures of juices and ethanol that are garnished with fruit and decorated with little straw hats. A few who are impervious to the evidence about UV-induced skin damage and cancer frequent tanning parlours when they are trapped in cooler places and, like the tanned majority of my early days, do everything possible to transform their delicate, outer integument into the golden brown that still symbolises health, wealth and happiness to many. Others, like me, just don't have the right genetic heritage to participate fully in this sun-baking culture.

My hometown of Brisbane is roughly the same distance from the equator as Miami, Johannesburg and New Delhi. At least early on in the brief, 220-year history of European Australia, an English/Scots, Anglican/Presbyterian establishment tended to dominate the cooler, more southern cities of Sydney, Melbourne and Adelaide, the latter being the only one not to have taken convicts. Starting as the Moreton Bay penal colony, the less desirable and (before air conditioning) more climatically oppressive Brisbane became the new home for large numbers of fair complexioned Irish Catholics, a 'Celtic fringe' effect that can still be seen in the professional life of the city. That transition from the mild dampness of the Emerald Isle to the heat and unreliable rainfall of south Queensland must have been quite a shock. The compensation was that, freed from overt political repression, many seized the opportunity to build highly successful lives.

The notable Irish nationalist and Dublin medical practitioner Kevin Izod O'Doherty (no relation), for example, became involved with the Young Ireland movement and, as co-editor of the Nationalist *Tribune*, was transported to Australia as a dangerous radical in 1848. From 1867 to 1885 he lived in Brisbane, held various elected roles in the parliament of the Crown Colony of Queensland and had eight children. Like the convict Magwitch in Charles Dickens's *Great Expectations*, he was unable to resist the nostalgic pull of 'home' and, returning to Ireland, stood successfully for election and sat as the member for North Meath in London's House of Commons. But, more fortunate than Abel Magwitch, he survived this experience of 'the expensive cure'. Kevin Izod came back to Brisbane and lived till 1905, long enough to see the 1901 Federation of States that formed the Commonwealth of Australia.

The high numbers of immigrants with clear, pale complexions that had evolved to let enough UV through the skin to enable vitamin D synthesis (from cholesterol) in the misty, wet climate of their native Ireland is just one of the factors that led to Brisbane being the melanoma and skin cancer capital of the world. Another factor is that, having a lively and perceptive medical establishment, the locals quickly set up substantial clinical and research programs focused on the problem. Disease incidence statistics are always related to levels of awareness and reporting.

The White Australia Policy that prevailed from 1901 to 1973 was an initiative driven at first by workers and trade unionists to prevent the importation of poorly paid, dark-skinned, indentured labourers from the Melanesian and Polynesian islands to the immediate tropical north. The result was that many who slaved all day in the blazing Queensland sun were grossly ill-equipped in the biological sense to deal with such high levels of solar exposure. Living later in the equally oppressive summer heat of the American South, we were surprised by the prevalence of white-skinned, auburn or sandy blonde-haired women with flawless complexions that seemed untouched by the sun. They were descendants of the pale, Protestant Scots/Irish who ran many of the cotton plantations in the Mississippi Delta, and the difference was that their culture had historically depended on black Africans to do the outside work. Being dark skinned, or burned by the sun, was not socially advantageous then for a Southern woman, though it can now be a political asset in cities like Memphis, which has a large African American population. Also, the beach-oriented lifestyle that so dominates Australia has never figured prominently in that part of the world.

As EM Forster pointed out, an excess of solar exposure ensures that the so-called white man in the tropics is more often the 'pinko–grey' man, though whisky may also have a bit to do with that. Perhaps to stave off the damp cold of peat bogs and the like, the Celts also evolved (culturally at least)

58

to be notable imbibers of spiritous liquors made from barley and other grains. After World War II the Australian government actively solicited grappa and wine-drinking immigrants from Greece and Italy, part of the idea being that much more appropriately pigmented (though still 'white') peoples from the Mediterranean and Adriatic shores would provide just the right workforce for the cane fields of tropical north Queensland. Some stayed in the tropics but, even before mechanical cane cutting became the mode, Melbourne—1370 kilometres further from the equator than Brisbane—had become the world's third-largest Greek city, and Melbourne's Italian-dominated cafe strip in Carlton was well established.

Well-presented, publicly funded propaganda has ensured that all Australians are now very aware of just how much damage the sun can do. Little kids wear hats to school, and the 'slip-slop-slap' mantra has us slathering various types of sunscreens on the exposed parts of our bodies. The early products used para-aminobenzoic acid ($C_7H_7NO_2$), but more recent variants are available to circumvent the allergy problems that can result from high-level skin exposure to compounds (like PABA) with benzene ring structures. These chemical sunscreens protect by absorbing the incoming UV radiation. The various cancer societies throughout Australia market swim gear that limits the extent of UV-induced skin damage. It's intriguing that acceptable beach wear for those of us who are not naturally armed against the sun's rays has gone from neck-to-knee to bikinis and ultra-abbreviated male Speedos and back to full cover again in less than a hundred years.

The experience that every Australian kid still has of being burnt by the sun is through the soles of the feet while running across one of those long, white stretches of beach sand to reach the breaking, cooling waves in the height of summer. Not far from Brisbane is Surfers Paradise, on Queensland's south coast, now a massive Miami-like resort with the magnificent beach often shaded by high-rise apartment blocks. My first memory of Surfers is as a beach locale boasting a single hotel with an attached, rather decrepit, zoo and a few, primitive holiday shacks. As it grew through the years, Surfers became the focus for school and university post-exam 'raves' in the heat of summer. One great character widely remembered from that era is the bronzed-beyond-belief guy who dispensed mutton-bird oil from an antique Rolls Royce parked on the esplanade. Apart from a baste-and-bake effect, a spray of mutton-bird oil does little for the skin—let alone block UV. Then there were the gloriously tanned meter maids (buxom young women who fed coins into parking meters); they were much admired at the time but may now regret getting a bit too much sun in their youth.

Though my English-origin mother, Linda Byford, had a dark, well-pigmented olive skin, I inherited the blight of fairness and freckles from my Anglo–Irish father. We went for beach holidays in the middle of winter, but I still got sunburnt. About the only effective sunscreen available at that time was the opaque, white, zinc cream (titanium oxide). Zinc cream was generally smeared only on the nose and ears. It picked up sand and, used in excess, made the wearer resemble the mime artist Marcel Marceau.

My dad was very proud that he had managed to come by the formula for a product called 'Surf Tint'. This had been taken off the market, but he replicated it, making it up from (as I recall) a mixture of methylated spirits, Friar's balsam and shellac. Perhaps I've got the formula wrong, because that also sounds a bit like one of the many recipes for French polish, the traditional finish used as late as the first half of the twentieth century for high-quality furniture. Eric Chippendale Doherty was a skilled amateur woodworker but I expect that, if it was a dual-use product, he varied the concentrations of the various ingredients according to the intended target. Slapping on 'Surf Tint' turned the user an acceptable shade of brown, the meth spirit gave some cooling effect as it evaporated into the atmosphere, then the shellac tightened the skin to give a sensation of varnished stiffness. If it had any UV protective effect, which I have good reason to doubt, it was probably due to something in the Friar's balsam—which, according to Webster's dictionary is 'a stimulating application for wounds and ulcers, being an alcoholic solution of benzoin, styrax, tolu balsam, and aloes.' Friar's balsam continues to find favour in the linked worlds of 'natural' medicine and aromatherapy. Whether 'Surf Tint' was withdrawn from sale because it didn't work, because applying a lot of Friar's balsam to the skin caused allergies, or because of methanol associated toxicity, the experience of using it as a child left me with a profound distrust of homegrown remedies.

After numerous episodes of sunburn, I soon became convinced that the safest practice was to stay in the deep shade. Even under a beach umbrella, 20 per cent of incident UV is reflected back from the bright sand. Though I loved, and still love, being beside the sea, I learned to value walking on the beach in the early mornings and evenings, and to delight in the play of moonlight on breaking waves while surf-fishing. Even so, in the culture of that time, the sun couldn't be avoided completely. Summer exposure on a cricket outfield or tennis court left me dehydrated and more than a little toxic from the resultant skin damage. I was delighted when I discovered squash, which is played indoors, and snow skiing, where it is easy to cover, or protect, most of the body's surface.

Apart from Alan Sherman's insight that 'you gotta have skin ... it's the thing that keeps you in' (which isn't accurate as that's as much the function of muscle and various types of connective tissue), how does the skin work? Consisting of the superficial, more 'armoured' (by keratin) epidermis and the dermis below, the skin is the largest organ of the body and our major interface with the dangers posed by the external environment. Made by the keratinocytes that form the protective 'stratum corneum' surface of the human epidermis, the fibrous structural protein keratin is also the basic component of hair, hooves, horns, feathers and the protective plates of the crocodile.

Heat and cold are sensed through direct effects on specialised, sensory nerve endings in the dermis and other 'exposed' (to air temperature) surfaces like the lining (mucosa) of the nose. These nerve endings are at the tip of long extensions (or axons) emanating from neurons (nerve cells) in sites like the trigeminal ganglia (which serve the face), or the spinal cord dorsal root ganglia (which receive input from much of the rest of the body). The ganglia are like relay stations: they facilitate the transmission of input sensory signals so that the right message transits to the 'effector', or motor side of the brain.

When triggered, the motor neurons forward their instructions via axons that innervate the muscles and thus initiate movement. Appropriate protective measures can be mediated via the subconscious, or autonomic, nervous system that (for instance) sends signals to cause relaxation of the involuntary, 'smooth' muscle that lines blood vessel walls (vasodilation) to radiate excess heat, or contraction (vasoconstriction) to conserve body temperature when cold. In hot weather, other signals will activate the sweat glands in the skin to secrete the moisture that provides the cooling effect when heat energy is given off during the process of evaporation. On the other hand, cold can induce shivering, the spasmodic muscle contractions that burn blood oxygen and glucose to produce heat.

The connection between the sensory side and the motor side of the brain can operate via a simple reflex arc; this registers, for example, as a sharp kick when our GP taps below the knee cap (patella) with a hammer, or when our hand is quickly withdrawn because it's too close to a hot flame. Other responses that require a more complex, integrated decision process concerning heat and cold are forwarded up the nerve fibres in the spine to the brain. Thanks to further neuronal activity triggered by signals within the brain, together with additional input from the eyes, we may use voluntary arm, hand and leg muscles to either put on a hat or move out of the sun.

If the ambient temperature strays too far from our normal 37 degrees Celsius (below 8 degrees or over 45 degrees), a sudden experience of heat or cold registers initially as pain. The sense of pain can, however, be quickly diminished by the production of endogenous brain opioids (like endorphins) so that we are able to enjoy the comforting benefit of a not too hot or cold compress. The application of heat can provide more blood, oxygen and nutrients to a strained muscle, while cold will decrease the blood flow to slow various chemical reactions that stimulate the pain nerve endings and contribute to the agony of severe sunburn.

How do the photons in sunlight cause the skin damage that leads to the reddening, pain and sloughing (peeling) of dead skin that we experience acutely as sunburn, then the delayed effect that can manifest as cancers like melanoma and squamous or basal cell carcinoma? The culprit is the UV part of the spectrum: the shorter the wavelength, the greater the potential for serious, long-term harm. In the laboratory, we use UV lamps that emit in the ultra-short UV-C range for sterilisation. These UV-C rays quickly destroy microbial DNA after first penetrating the outer coat of viruses, bacteria and fungi. Once the DNA is massively disrupted, the micro-organism dies. A smaller UV dose can lead to non-fatal DNA breaks (mutation) and, if the target is a skin epithelial cell (forming the body's outer tissue layer) rather than some sort of infectious bug, the possibility of cancer. Fortunately for us, all the UV-C emitted by the sun is blocked by the earth's atmosphere.

Some of the medium-length UV-B, most of which is taken out by the stratosphere's ozone layer, does get through, particularly when the noonday sun is right overhead. In the evening and early morning, the UV-B is refracted away from the earth's surface. Incident UV-B has long been considered the prime cause of radiation-induced DNA damage that can, in turn, lead to the division of viable, mutant cells and eventual escape from growth control—which is the disease known as cancer. More recent evidence is also implicating the long wave band, UV-A, which is the predominant component of the solar UV spectrum that reaches us. The eventual consequence of UV-induced DNA sequence modification can be either the relatively harmless (unless they change further) scaly hyperkeratoses or, much worse, the smooth, shiny bumps that are typical of a melanoma or a carcinoma.

Exposure to UV-B has the positive effect of inducing highly specialised cells called melanocytes to increase the synthesis of the complex polymer melanin. The melanin accumulates as granules that are then exported via channels called dendrites to the neighbouring keratinocytes. There the dense melanin granules sit like mini sunshades over the DNA-containing nuclei and protect the keratinocytes from radiation damage as they move upwards to form the tough, continuously dying and desquamating (flaking off) stratum corneum.

The dynamic process of development from specialised stem cells, movement to exposed surfaces then loss that characterises the keratinocyte life cycle is also normal for other surface epithelia like those in the gastrointestinal tract. The progress from birth to destruction that is so aptly summarised in the Anglican Book of Common Prayer burial service—'in the midst of life we are in death'—is central to the good health of all biological ecosystems. The cells of complex life forms are programmed to commit 'altruistic' suicide if they go wrong in some way. The scaly, surface layer of the epidermal stratum corneum is largely comprised of dead cells and, even if they are still alive when they encounter some kind of insult, the keratinocytes don't hang around to create further problems for the organism at large. If, however, random, UV-induced mutation causes changes to nuclear DNA that make a keratinocyte and its progeny cells immortal, then we have the precondition for cancer. As with human beings who would quickly crowd out the planet, immortality is not a good idea for most body cell types. The exception is the neurons that don't normally divide subsequent to the earliest stages of life.

Skin melanin comes in two basic forms: the black or brown eumelanin that determines the colouring of dark-skinned people and those who tan well, and the red phaeomelanin made by the 'pale faces' among us. Freckles are due to the spotty production of eumelanin. Melanocytes located in the hair follicles are also responsible for hair colour. Black eumelanin predominates in brunettes and phaeomelanin in redheads. Blonde hair is due to the presence of a fairly small amount of brown eumelanin, while the greying we associate with aging and our tropical pinko–grey man reflects the time-related loss of the hair follicle melanocytes.

The more fortunate of the fairer members of the human race, including many whose ancestors lived in periodically sunny areas like the Netherlands or south Germany, have an ample supply of melanocytes programmed to produce brown eumelanin. Relatively pale through the dark days of winter to allow UV-induced vitamin D synthesis, the increased light intensity of spring induces the production and oxidation to the darker form that sees them through the solar challenge of summer. Dark-skinned people, who have evolved in areas where intense sunlight is the rule, maintain high levels of black eumelanin year round. As a consequence, they suffer much less from the long-term wrinkling and 'crocodilean' transformation that is such a prominent feature of UV-induced skin aging in northern Europeans who, departing from their ancestral homelands, have lived more of an outdoor life in hotter climates like California or Queensland.

The reddening and pain of sunburn reflects the body's response to heat and the molecular damage caused by UV-A and UV-B. Modern sunscreens are designed to block both wavelengths. Injured, or dying, skin cells release various specialised chemicals (with names like leukotrienes or chemokines), some of which function to recruit other scavenger cells (the monocyte/macrophages) from the blood. These phagocytes (big eaters) help to clear the cellular debris and secrete a range of molecules that, while contributing to the level of physical discomfort sensed via nerve endings, act to promote tissue repair. With time and healing, the acute damage and pain resolve, but the accumulation of connective tissue, collagen and the like leaves permanent scars that contribute to later wrinkling and 'crocodile' skin.

As with any manifestation of a serious deity, the Sun God has the capacity to confer great benefit on the one hand and major harm on the other. Also, like most divinities, the Sun God seems pretty much indifferent to our fate. A world without the warmth and light of the sun would soon become cold and dead, but we can have too much of a good thing. The accumulating evidence that ever-increasing concentrations of atmospheric CO_2 will cause the sun's rays to heat the planet to levels that are dangerous for many life forms (including us) can no more be ignored than the sure and certain knowledge that exposure to excess UV will lead to sunburn, permanent skin damage and even cancer.

Advances in technology and informed changes in behaviour over the past twenty years have greatly mitigated the toll taken by the sun on the more vulnerable members of our species. When threatened with consequences like sunburn and skin cancer, individual human beings find it relatively easy both to adopt the appropriate countermeasures and to accept that there will be some associated financial costs. Sunscreens, protective swimwear, broad-brimmed hats and beach tents sell very well in the Australian summer.

We need to build on those linked senses of personal vulnerability and responsibility as we seek to promote collective approaches for dealing with the much more difficult issues of energy conservation and the need to replace CO_2-producing fossil fuels with renewable, non-polluting sources of heat, light and power.

Burnt By the Sun, the intriguing 1994 Russian movie, wasn't about sunburn or global warming but dealt with 1930s paranoia and repression under Russia's 'man of steel', Joseph Stalin. Any idea that some form of rigid, centralised, authoritarian control is what we need in order to deal with the greenhouse gas problem is quickly laid to rest by recalling the appalling air, water, land and nuclear pollution caused by the old Soviet Union. In the end, reality exposes those who abuse power and promulgate deliberate lies. However, embracing laissez-faire attitudes that, from a mix of intellectual and political laziness, cede control to unscrupulous, private vested interests, will also lead to disaster. Capitalism that is unrestrained by human compassion or thought for the future is, in the long run, just as bad as totalitarianism—or even worse when it is globalised and not subject to meaningful regulation.

We can't afford to let the infinitely greedy, corrupt and/or delusional among us lead humanity down the path encapsulated in Louis XV of France's cynical statement, '*après nous la déluge*' (after us the flood). The mix of manipulation, flamboyance and triviality that ensured political stability under the long (1643–1715) rule of Le Roi Soleil, the Sun King Louis XIV, served only to conceal destructive, unresolved tensions. The waste of the national treasure on useless wars and the inability to deal with the needs of normal people persisted through the reign of the much less competent Louis XV. It took fewer than eighty years to move from the cake and circuses of Louis XIV's court at Versailles to the 1789 French Revolution, citizen Robespierre and the subsequent Terror.

By flicking a light switch, starting a car or flying in a Boeing or an Airbus, the humblest among us take steps that the Sun King could not even have contemplated. Our era is infinitely more brilliant in the technological sense than anything experienced at the glittering, mirrored Palace of Versailles by eighteenth-century aristocrats and their servants. The inexorable, revolutionary, corrective force that we face, though, is not the anger of the *sans culottes*, the disenfranchised and the poor, but the constraints imposed by nature itself. The problem for us is to deal with the consequences of human inventiveness, on the one hand, and unthinking profligacy on the other. Finding solutions will test the outer limits of our self-discipline, intellectual capacity, practical ingenuity and moral integrity. Failure to act will be judged very harshly by future generations that burn in the sun.

Iron Horses and
Balladeers

On steam

Herbert Lawrence (Bert) Byford, my
maternal grandfather, was a railway man from the age of
steam. Born in 1875 in a mill on the banks of the river
Chelmer at Little Baddow, near Chelmsford, Essex, he
entered a world where the Industrial Revolution was trans-
forming both air quality and the British way of life. Steam
trains were replacing horse-drawn canal barges as a means of
transporting goods across the English countryside. Steam
rather than wind or flowing water was the force driving
wheels in mills of all types and descriptions. Even in the quiet
countryside, steam traction engines moved from farm to farm
to power the belt-driven threshing machines that separated
the grain from the chaff. Lovingly restored and brightly
painted, these cumbersome mechanical dinosaurs, with their
steam whistles and spinning flywheels, are often to be seen
today at agricultural fairs and the like.

By 1850, Little Baddow's most famous son, England's longest serving (1829–74) hangman, William Calcraft, could travel some six thousand miles (9660 kilometres) of railway tracks as he flogged and strung up miscreants all over the country. In 1865, for example, Calcraft despatched Dr Edward William Pritchard in Jail Square, Glasgow, condemned for the murder of his wife and mother-in-law. Calcraft's performance (he was master of the 'short drop' that did not kill too quickly) drew a crowd of 100 000. Comparing that with the 51 400 capacity of Glasgow's Ibrox football stadium, where the hotly contested Rangers (Catholic)/Celtic (Protestant) soccer games are played, does raise the possibility that changes in public entertainment and 'values' over the past 150 years or so have not all been for the bad.

A kind, gentle man, my grandfather was the polar opposite of his illustrious village elder William Calcraft. It's unlikely that Bert Byford strayed far from home until he boarded the modest twin-screw, single-funnel, Aberdeen Line SS *Marathon* in 1903 and sailed across the world, via Cape Town, South Africa, to Melbourne, Australia. Arriving in August, he then travelled north to Brisbane, probably by rail, to marry his fiancée Emma Eliza, the daughter of Chelmsford gentleman farmer and grain merchant Peter Bedford Smith, who had emigrated during the previous year to join his brother in the timber business.

Bert was already well established in the railways by the time he left England as a twenty-eight year old. We don't know the details but, being from the Chelmsford region, it's likely that he worked for the Great Eastern Railway that, in 1923, amalgamated with parts of the Midland and Great

Northern Railway to form the London North Eastern Railway, the iconic (to train buffs) LNER. Since 1862, the Great Northern Railway's *Scotsman*, later to become the LNER's *Flying Scotsman*, has been departing London's King's Cross (platform 10 at 10 a.m.) each day for Edinburgh's Waverley Station. When I took a job in Scotland, and thus became the first of Bert and Emma's descendants to return to the United Kingdom, we sailed by ship to Southampton, stayed in London for a couple of days and completed the journey to Edinburgh on the *Flying Scotsman*. Having very little money at the time, we still recall the extravagance of eating salmon for lunch in the dining car, a service that was added in 1900.

By 1967 when we made our trip, the train was drawn by a diesel/electric Delta locomotive, replacing steam engines like the massive, streamlined Mallard that, clocked on a slightly downhill grade, hit 126 miles per hour (more than 200 kilometres per hour). At close quarters, there's nothing more dramatic than the bulk and momentum of an express at high speed. Remember the scene from the film *Dr Zhivago* where, belching steam and smoke, with red flags of the revolution flying, the armoured train commanded by Strelnikov (based on Leon Trotsky) rushes past the sidelined passenger transport carrying the physician Yuri and his family? Now the whole 627 kilometres of the Great North Eastern Railway's *Flying Scotsman* route is electrified, with the time for the journey being cut from more than seven hours in 1938 to fewer than five hours today.

In 1925 the *Flying Scotsman* was the first train to exceed 160 kilometres an hour, while the earlier Atlantic-class engines that operated before my grandfather left England could shift her along at an average speed of 96 to 112 km/h and up to 128 km/h, beyond the legal highway limit that currently applies in many countries, including Australia. That compares with the first railway out of London, Robert Stephenson's line to Birmingham, which opened in 1837 and by 1844 had reduced the 182 kilometres from between Euston, London, and Moore Street, Birmingham, to a four-hour journey. Even then, though, the engines could travel safely at speeds in excess of 80 km/h, marking the beginning of the era when people moved across the surface of the earth faster than a galloping horse and lived to tell the tale. The contrast with jolting around in a poorly sprung, horse-drawn coach on bad roads must have seemed like a dream transformation.

When Bert Byford disembarked from the *Marathon*, he was landing in the first Australian city to build a steam railroad. This ran from Melbourne to Port Melbourne and was completed in 1854. By 1904 all the Australian States were criss-crossed with railway lines that moved people, and brought produce from the hinterland to the coastal, port cities. Unfortunately from the viewpoint of uniformity, these networks were established before the 1901 Federation that formed the new nation of Australia. The decisions made back then show how little real co-operation there was between the six Crown Colonies isolated on this remote continent. The three eastern colonies all chose different rail gauges: Victoria 5 feet 3 inches (1600 mm), New South Wales 4 feet 8½ inches (1435 mm) and Queensland 3 feet 6 inches (1067 mm). This choice of wide, standard and narrow gauges, was a disaster for interstate trade—a problem that was only partially corrected with the completion of standard gauge lines connecting all the Australian state capitals late in the twentieth century and the 2004 opening of the Adelaide to Darwin link. If Bert travelled from Melbourne to Brisbane by rail in 1904, he would have had to change trains at Albury on the Victoria–NSW border, then again at Wallangarra on the NSW–Queensland border.

The same problem arose in the United States, where privately constructed and financed railroads were the national model. That difficulty was largely resolved in 1886 with the adoption of the international standard gauge used in the North Eastern corridor by the prominent Pennsylvania Railroad. The net result is that freight cars move freely across the USA. One way to make money today is to buy a near-defunct short-haul railroad, then put the rolling stock into circulation on the national grid. This makes for very variegated freight trains. Memphis is a major east–west, north–south meeting point and distribution centre, with several railroads running through the city and suburbs. That was already the reality when this strategically important southern city was, after a naval battle in the Mississippi, occupied in June 1862 by the Union's Western Army commanded by Ulysses S Grant.

Memphis is very flat, so there are few opportunities to build the bridges that often carry roads across railway cuttings. As a consequence, streets are frequently blocked when the red lights flash and the boom gates come down to prevent that meeting between a locomotive and a truck or car that is generally fatal for the occupants of the smaller vehicle. Every year, someone from around Memphis is killed on an un-gated crossing or, in some cases when, often alcohol-fuelled, they try to beat the train and duck around the barrier. They take the chance because those immensely long US freights can take forever to pass in front of your windscreen.

A couple of enormous, throbbing diesel locomotives might be hauling a mile of coal wagons headed for a power station or a steel plant. Sometimes there's an added diesel at the back, pushing to provide extra oomph. Despite the concerns about global warming, there seem to be more and more coal wagons being hauled all over the planet. In recent times, we may be reminded of our bright, new globalised world by trains of special flat-bed trucks, each carrying two of those big shipping containers bearing names like Maersk, Hapag/Lloyd, Tex, or Hyundai. Other trains are collections of boxcars labelled Burlington Northern Santa Fe, Illinois Central, CSX, and Norfolk Southern, together with lesser known brands. The word linkages conjure up all sorts of romantic images relating to place and time and recall earlier combinations like Erie/Lakawanna, Wabash & Michigan, Yazoo & Mississippi, Baltimore & Ohio and the Union Pacific and Central Pacific, which (in 1869) met in Promontory, Utah, to connect the east and west coasts of the United States.

As the freight cars bump by while we wait for the boom gates to lift we might—if the air-conditioning is on, the car windows are closed and we have that Southern sense of not being in too much of a hurry—even launch into one of those great railroad songs. Memory is a strange kind of storage system and my brain, like many brains I suspect, retains the odd set of words and verses from ballads like *Freight Train,*

Freight Train, Going So Fast. That works fine if the train is moving at speed, though the progress often seems snail-like, particularly in urban areas. Then there's *John Henry was a steel driving man*; *The Atchison, Topeka and the Santa Fe*; *The Ballad of Casey Jones*; *The Wabash Cannonball*; *The Chattanooga Choo Choo*; and, of course, Roger Miller's unforgettable 1936 'hobo' depression ballad, *King of the Road*: 'Third boxcar, midnight train, Destination … Bangor, Maine.' Another tune stuck in my head is the old college drinking song: 'Riding down from Bangor, on an eastern train, after weeks of hunting, in the woods of Maine'. 'Train' and 'Maine' are pretty much irresistible to a railway lyricist, and the name 'Bangor' has such a great euphonic ring to it.

If we exhaust our few remembered verses, there's always: 'Passengers will please refrain from passing water in the train while it's standing in the station yard … if you feel you really oughter kindly call a Pullman porter', sung to the tune of *Humoresque*: dum de dum … de dum de dum … de dum de dum de dum de dum … and so on. Variants of the first part of that message were in every railcar toilet throughout the world in the not so long-gone days when the pedestal simply emptied to the track beneath. At least for a male, one of the small dramas on a long train trip was balancing in the swaying cubicle and aiming at the stones rushing by on the rail-bed below. All the great railway songs I know are from the United States, and they have a lot to say about the lives of working men in the mid-nineteenth to mid-twentieth centuries. The Pullman porters (sleeping car attendants) were adult, mostly black males, as would have been the shoeshine 'boy' in 'Pardon me Boy, is that the Chattanooga Choo

Choo? Track 29, Boy, you can give me a shine?' The real-life Casey Jones, the skilled and brave *Cannonball* engineer, was white, while the hard-working fireman Sim Webb, whose life was saved by Casey, was black.

Arriving in Brisbane in 1903, Bert Byford found work in the railways, after a couple of interim jobs, and was soon promoted to signalman, the role he had been trained for in England. The young family was sent for a time to Maryborough, the home of Walker's Locomotive Works, which, beginning in the 1870s, made the green-painted mainline locomotives of my childhood and which still functions today. They then moved to the Brisbane suburb of Clayfield, where they were flooded out of their house. Their final relocation was to the three-bedroom, wooden 'worker's cottage' in the outer suburb of Oxley where Bert and Emma raised four children and lived until well into their ninth decade.

At the end of a dirt track, the little house was across the street from the main line that went west from Brisbane to Charleville. The first time I left home for more than a couple of weeks was when I boarded the diesel/electric-hauled *Westlander* to spend a three-month university vacation working at Cunnamulla, in the dry, open, mulga (*Acacia aneura*) country 750 kilometres inland. This western line was also a first for the state of Queensland. The initial stage stopped at the nearby Bigges Camp (now Grandchester) in 1865, then it was progressively extended until it finally reached Cunnamulla in 1898 via a spur that came down from Charleville.

The view from my grandparents' front veranda was to an embankment with a wooden hopper and, sometimes, a coal wagon sitting on a little rail siding. The coal, released via a chute, was taken by motor truck to the local pig abattoir, known colloquially as 'the bacon factory'. Steam, smoke and hot water all play a part in the post-mortem processing of pigs for human consumption. Bert supervised the coal transfer activity, but his main job was switching signals and points in the Oxley signal box, which he reached by walking about 250 metres. If you don't know what a signal box is, think back to (or rent) the movie *Von Ryan's Express*, where the prison camp escapees led by Frank Sinatra enter an elevated wooden cabin by the tracks and encounter the waist-high switch levers that shift trains from one line to another. (You can't steer a train; it goes where the tracks take it.) My grandfather was proud of his job and his responsibilities. He worked hard, put in a lot of overtime and, as people did at that time, retired at age sixty-five and took the government pension.

The signalmen in their trackside signal box were the historic equivalent of air traffic controllers in the tower of a modern airport. The difference was that, although they communicated with colleagues up and down the line by tele-graph and later telephone, their messages to the train crews were relayed by positioning the pivoting blades of those big semaphore signals in various configurations and by showing lights of different colours. Railwaymen had to learn a whole 'knowledge' of signals. Some conveyed immediate commands like STOP, SLOW, GO, while others gave information on the longer distance situation. A horizontal blade plus a red light always meant STOP, while a dropped blade and green was GO, but there were many variations in colour and the shapes and sizes of the signal blades.

Get the signals and the switches wrong, and the result could be a major train wreck. *The Ballad of Casey Jones* celebrates how the Illinois Central engine driver sacrificed himself by staying with the engine and hauling on the brake to save his fireman and passengers when the *Cannonball*, rushing south from Memphis to New Orleans, drove into the back of a sidelined freight that hadn't cleared the main track in Vaughan, Mississippi. Train wrecks are always dramatic. Did that haunt my grandfather as a railwayman who was responsible for preventing such occurrences? He cut pictures and stories of railway disasters out of the newspapers and kept them in a scrapbook. The Oxley signal box where he spent most of his working life was pulled down before I was old enough to remember, perhaps with a change from all mechanical to more electrical systems, and the switch levers that worked the points were transferred back to the single, island, passenger platform with its combined station-master's/ticket office and waiting room. The most unusual feature of the Brisbane railway stations of my early child-hood was that they were partly taken up by massive, concrete-block air-raid shelters, which were eventually demolished years after World War II had ended.

Through my high school years, I spent quite a deal of my life dreaming away on that railroad platform while waiting for the train that took me and my fellow students to the much more upmarket suburb of Indooroopilly. Oxley was where the 'local', known colloquially as the 'turnback', started its 'all stations' journey, passing through the suburbs of Chelmer and Indooroopilly (among others) to the city. Pulled by a dark blue 'Thomas the Tank Engine', our slow train was sometimes delayed because the express carrying commuters from the coal-mining town of Ipswich to the commercial heart of Brisbane was late. Then the long, green, mainline locomotive hauling its string of carriages would clatter past, relieving the boredom of railway limbo and releasing us to begin our more modest progress.

We take the familiar for granted. During all that time spent hanging around waiting for the signal for us to board, I thought very little about the compact, steam-breathing unit of iron, fuel, fire and water that would soon lead us down the polished metal tracks. I was aware that the driver and fireman were standing (or perched on little seats) in the cab behind the horizontal, riveted barrel of the engine's boiler. The roles of the headlamp and the smoking funnel (smokestack) were obvious and it was impossible to ignore the shrill shriek of the steam whistle and the occasional release of high-pressure steam from the safety valve. It was only when I looked up how these iron monsters actually work that I sorted out, among other things, what those big domes on top of the boiler are for: one collects the steam, while another carries sand to drop on the rails when the wheels need more traction. Then the outside of the engine was festooned with various bolted-on, steam-driven, hissing, clunking gadgets, like the cylinders with their pistons and connecting rods, the electric generator, the compressor for the air brakes, and the water pump.

Sometimes the fireman would open the door of the firebox, and then we could see the flames from the burning coal. How does such a little fire that is behind, rather than under, a large container of water bring it to the boil? The way it works is that the boiler is honeycombed with narrow, horizontal tubes that convey the heat and smoke from the firebox through to the smoke box, then out through the funnel. If you've ever seen a photograph of a steam locomotive with that round door at the front open, what you're looking into is the smoke box with an end-on view of the water-filled boiler and its plethora of penetrating fire tubes.

The steam first collects in the steam dome and is then heated further by being passed back through the smoke box and boiler. Sent in a controlled way to the cylinders by the engineer (train driver) as he operates the throttle, the superheated steam forces the pistons down to move the connecting rods that link to the big driving wheels. Most locomotives have a single cylinder/piston on either side, each of which goes through two cycles of intake and exhaust in one turn of the drivers. The boiler, steam dome, steam tubes and cylinders are under high pressure and have to be constructed very strongly to avoid the possibility of an explosion. That can become a reality if the water level falls too low, which is why the water pump that maintains the supply is very important.

Larger, industrial applications, like powerhouses, use water-tube rather than fire-tube boilers, avoiding the inherent danger of what is effectively a large, almost completely sealed, pressurised kettle by limiting the volume of boiling water to that carried in the tubes. Later versions of steam automobiles had water-tube boilers that allowed them to generate a sufficient head of steam for them to move off in forty seconds, though the famously fast Stanley Steamer (1902–27) had a fire-tube boiler. Steam cars were effectively driven out of business by the invention of the electric starter motor, which meant that potential purchasers no longer needed to choose between waiting till a car got up steam or risking a broken arm from cranking a heavy internal combustion engine. That latter experience would no doubt cause the suffering 'automobilist' to 'let off steam' in a dramatic and loud way.

Unlike the green mainline locomotives that dragged their coal and water behind in a separate tender, our blue Thomas the Tank Engine had a protruding hopper full of coal behind the cab and the water tanks up front alongside the boiler. This meant that Thomas could be driven equally well forward or backward. When the 'turnback' that I caught to go to school reached the end of its run at Oxley, the crew would unhitch the coupling to the first carriage and disconnect the high-pressure airline that served the brakes. Then, with the appropriate switching manoeuvres on the part of the signalman, the locomotive would zip around a parallel track to be connected up again and head the train back to the city centre. The engineer and fireman might also have topped up their water and coal supply as they completed their bypass manoeuvre. A 180-degree change in direction for an engine plus tender, on the other hand, requires either a turntable or a large loop line.

The long-distance trains of that era had the side-aisle coaches and compartments that will be familiar from watching movies like *Murder on the Orient Express* or from travelling in modern sleeping cars. Constructed of varnished red cedar and other local timbers, they were decorated with fine black-and-white photographs depicting scenes from rural Queensland. While the 'first class' passengers on the commuter trains might have enjoyed the central corridors we all know today, the 'second class' carriages were 'toast racks'. The passengers sat staring across at each other on parallel, long, bench seats that stretched the width of the carriage. Accessed by a door at each end, these mini-compartments were open above shoulder height throughout the carriage, making it pretty hard for the adolescent school kids that rode the trains in the mornings and afternoons to get up to much in the way of 'hankey panky'. To further discourage such behaviour, both sexes wore generally drab uniforms of unrelenting modesty, often with neckties in the heat of the Queensland summer. I don't recall seeing any 'valley girls' on the 'turnback' and, in any case, the 'likely lads' of those days would have been totally intimidated by such exotic and incomprehensible creatures. If the boys and girls communicated at all, they tended to use words and whole sentences.

Both the little blue tank engines and the beautifully constructed wooden carriages were built and serviced at the Ipswich Railway Workshops. Looking down, though, while waiting for the engine to do its head-to-tail stuff, I recall that the axle wheel boxes all carried the stamp 'Timken Roller Bearings', which means that they were imported from Canton, Ohio. We couldn't board until given the nod by the guard who rode in a separate, enclosed compartment in the last carriage. The freight trains had a separate guard's van, or caboose, but, unlike some US configurations, it was always at the end. The train could only move off when the fireman had stoked and trimmed the furnace, the guard had shouted 'all aboard' and waved his red flag, and the driver had released the brakes and blown the steam whistle.

As with the black gangs on ships, the hardest working crew member was the fireman. An automatic stoking mechanism was used on some of the big, long-distance, twentieth-century US trains, as the effort required was just too much for one man. When Australia's World War II prime minister, the self-educated, former engine driver Ben Chifley, joined the failed strike against the New South Wales railways in 1917, he was fired then re-employed but demoted to fireman. The strike was a major confrontation that spread also to coal miners. Re-building the de-registered union helped Chifley develop a political constituency among working people and led to his parliamentary career in the Australian Labor Party.

My grandparents and their four children are now gone, as are the steam engines and wooden 'toast racks' from Brisbane's commuter rail network. First diesel, then electric, locomotives drawing metal carriages made the local trains look much the same as those anywhere else. Diesel/electric locomotives haul the long-distance expresses and freights. Queensland has a lot of high-quality coal, so it's easy to understand the changeover from steam engines to electricity derived from coal-fired plants. But oil supplies are limited, so why switch from steam to diesel? The first point is that oil is a much more concentrated and easily handled source of energy than coal. The fuel is delivered by a pipe feed to a fuel injection system, not by a fireman with a shovel, and there are no rock-like clinkers to clean out of a furnace. Diesel-hauled freight trains do not need to stop at regular intervals to take on coal and fresh water, and fewer people are required to maintain the more distantly spaced depots.

The second issue is that the action of two big pistons moving in and out as they turn the massive driving wheels at speed has an irregular pulse characteristic that is completely lacking for an electric motor, whether it's powered from an overhead pantograph or third rail collecting from a mains supply, or from a generator on board that's driven by a diesel engine. The transmission of the steam engine thumping action through the wheels to the rails is hard on the tracks, which require considerably more maintenance than is the case for a modern railroad. Adding a third, central high-pressure cylinder which then passed steam to the two, outer cylinders diminished this effect for the *Flying Scotsman*'s Mallard—as it did for some of the big German and US locomotives, but they came along fairly late in the age of steam and suffered the further penalty that they were immensely heavy.

But then, steam engines are, in just about every sense, very inefficient. Operating via the principle of external combustion, they burn coal to heat water that, in turn, gives high-pressure steam. It can take an hour or so to achieve a sufficient head of steam to pull a train, while getting a diesel/electric locomotive under way just depends on the equivalent of turning the ignition key in a car. Also, any internal combustion engine where the explosive fire (from burning oil) and the consequent heat-driven expansion process all occurs within the cylinder is much less wasteful of energy than a system requiring the transfer of steam from a boiler. Over the years efficiency was increased by modifying the valve gear, but too much energy still went up the blast pipe that exits via the funnel to produce the *choo-choo*, or *chuff chuff*, signature characteristic of steam rail locomotives.

Travel to Queenstown in New Zealand and you will hear something like the same sound from the TSS *Earnslaw* as the vintage 'puffer' steams across Lake Wakatipu.

At least for a time, steam-driven locomotives and steamships developed more or less in parallel. London's Science Museum, across the road from the Victoria and Albert Museum, has an extensive, permanent display showing the evolution from the earliest beam engines that pumped water from Cornish coal mines, through the industrial engines with massive flywheels that drove belts and spinning shafts to power multiple looms, to the steam turbines that sped the super-dreadnoughts and later battleships through the oceans.

The original, generally two-cylinder, horizontally opposed, low-pressure (no superheating) steam marine engines used in the first screw-driven warships (such as HMS *Warrior*) operated on a system that was not very different from that found in mills and on railroads. However, a ship floating in water can support a power plant that is much heavier than anything carried by a locomotive riding on a rail bed.

The preserved armoured cruiser/battleship USS *Olympia*, commissioned in 1892, has two three-cylinder triple-expansion engines, similar to the type that was still in use (though with a much longer stroke) in the mass-produced Liberty ships of World War II. These engines are the massive, vertical, iron piles that long survive the rusting hull plates of Liberty ships sunk by enemy action or sent to the bottom much later by those wanting to enrich a marine environment by forming an artificial reef. In a triple-expansion engine, the incoming high-pressure steam goes first into the smallest cylinder where it drives that piston down and loses pressure and increases in volume. Then it passes to the next-largest cylinder, and so on to the third, and biggest chamber. Finally, the steam exits through a condenser to be recovered as water, an approach that doesn't work readily for railway locomotives.

The largest of the three cylinders in each of the *Olympia*'s engines has a diameter (bore) of 234 centimetres. Think of the six- or eight-cylinder engine under the hood (bonnet) of a large family car. Then imagine a cylinder housing a piston wide enough for a 2-metre tall man to lie full stretch across with an added twenty or so centimetres at his head and feet. Operating together to drive the twin propellers, these engines pushed *Olympia* through the water at speeds of up to twenty knots.

By the end of the first decade of the twentieth century, steam turbines had become the propulsion system of choice for top-of-the-line warships and even passenger vessels, particularly the Atlantic Blue Riband Cunarders *Mauretania* and *Lusitania*. The 1911 RMS *Titanic* still had two triple expansion engines flanking a central steam turbine, reflecting the somewhat strained financial situation of the White Star Line. Parson's steam turbines drove the original (1913) HMAS *Sydney* at a top speed of 25.5 knots, while her unfortunate successor of the same name, a Leander class cruiser launched in 1931, was powered by a more advanced version of the same engines and could hit 32.5 knots. In a turbine, the incoming steam first strikes the smallest blades, then encounters larger and larger blades as the pressure drops. This is the principle used in a modern electricity powerhouse.

The steam powering the turbines can be generated from water-tube boilers using coal, oil, natural gas, nuclear or even solar energy as the heat source.

Steam turbine railroad engines were developed, but they came late and did not delay the transition to diesel/electric haulage. With associated tender, some of these behemoths were more than thirty metres long. The biggest locomotive that I've seen (twenty-seven metres) is the experimental Baldwin 60 000, the 357-tonne giant that is on permanent display at Philadelphia's venerable science museum the Franklin Institute, which bought it in 1933 for one US dollar. Completed in 1926, she was the sixty thousandth engine built by the Philadelphia's Baldwin locomotive works. The 60 000 was not a success: she was too heavy for the tracks and could carry only enough water to steam at full speed (112 kilometres per hour) for two hours. With some express locomotives that 'drying out' problem was solved by scooping water from a trench at the side of the rail bed, though this was a very inefficient process and the train had to slow to about 60 kilometres per hour to make it possible. Another difficulty with the Baldwin 60 000 was that the experiment of surrounding the firebox with water tubes did not work out, as they tended to burst.

Any process that uses high-pressure, scalding steam has inbuilt risks. In the past, when control systems were not as sophisticated as the electronic gadgetry available today, the steam autoclaves used extensively in hospitals and laboratories for disinfection and sterilising laboratory glassware, surgical instruments and so forth were a common cause of accidents and even fatalities. Steam fitters and those who operate even the smallest steam-powered vehicles, like some of the antique steam pinnaces that can be found on bays and lakes, must be both properly trained and certified. Too many things can go wrong. Disasters have been caused by blocked or improperly assembled safety valves while, as recently as 1977, nine people were killed and forty-five injured in Bitterfeld, Germany, when the water level fell too low and a locomotive boiler exploded. When that happens, hot spots cause the residual water, or added cold water, to boil off (flash) too quickly and create excessive, sudden pressure.

Low, or shifting, boiler water levels were also the likely cause of America's greatest maritime disaster when, in April 1865, one of the four fire-tube boilers on the Mississippi river paddle steamer *Sultana* exploded just north of Memphis. Licensed for 376 passengers, the *Sultana* was pushing upstream against the powerful Mississippi current with a cargo of 2400 Union soldiers who had just been released from Confederate prison camps. The death toll was estimated at around 1700, although the exact number isn't known. The worst such Australian disaster occurred in 1872 when the small side-wheeler *Providence* blew up on the Darling River: five people died and the bits and pieces were scattered for some 300 metres. Most of these riverboats were both constructed of wood and wood-fired, with the result that many more were also lost due to fire.

Apart from the safety aspect, coal-fired steamships and locomotives were dirty and required a lot of maintenance. In addition to the particle-laden smoke exiting the funnel, the boilers had to be cleaned regularly to remove the mineral scale that deposits in any device used to generate steam from ground water. Riding in a steam train with the windows open, you soon learned that it was best to sit facing backwards or at least to look away if the carriages were rounding a slow curve and the engine was in direct line of sight while pulling hard up a grade. Catching a large cinder right in the eye isn't an experience that anyone would want to repeat.

One of the continuing themes in my grandparents' house when the family gathered for Sunday lunch was how much coal dust and grime steam locomotives generate. 'Dirty old trains' was the operative phrase. 'Old' was a pejorative term in the 1950s, a dismal time in human history when otherwise sane people would cover the surface of antique, beautifully polished and aged woodwork with layers of pastel enamel. In retrospect, I suppose that this 'dirty old' steam train analysis was part of a gentle 'battle of the sexes' between Emma and their two daughters Frances and Linda (my mother) on the one hand, and poor old Bert the retired railwayman on the other. This was an era when most married women still stayed at home and did all the household cleaning. The smoke from the locomotives accelerating on tracks just across the street as they headed westwards from Oxley station indeed gave genuine cause for complaint.

As a small child I got drawn into this 'dirty old train' discourse, on the female side of course. I remember travelling alone with my grandfather on the train journey we took north each year to our annual vacation at the seaside resort of Caloundra. The rest of the family must have gone ahead, or maybe we couldn't all fit into the first car my father ever owned, a recently purchased 1934 Chevrolet Standard tourer. Automobiles were in short supply in the years immediately after World War II and, like many of his generation, Bert Byford never learned to drive. Now, with a new highway and suburban sprawl, Caloundra is almost an outer suburb of Brisbane.

We caught the local from Oxley to Roma Street station, boarded a long-distance train to disembark at Landsborough near the marvellous Glass House Mountains named by Captain Cook as he sailed up the coast in 1770, then caught a bus that took us to Caloundra via a feeder road from the main Pacific Highway. As I recall, this train trip with my grandfather was also the occasion of my first 'scientific' experiment. When the train stopped so that the passengers could refresh themselves with tea and cakes at Caboolture, he had bought me an ice cream, which came in a cardboard carton. After consuming the contents, I held the empty, moist carton in the open window of the moving train to see just how much soot it would collect. The strategy worked and I got a good yield, though I hadn't yet learned that you have to make accurate measurements and do replicates and controls if you want to reach valid conclusions.

Apart from the experience of having a railwayman as a grandfather and travelling on steam trains to school, steam also played a direct part in my childhood with the introduction of a gadget called the pressure cooker. The pressure cooker was a small, domestic autoclave used throughout Australian kitchens of the 1950s and 1960s to sterilise and cook meat and vegetables. The sealed container sat on a stove hot plate and hissed away as steam came out the heavy safety valve on top. As Australian farmers and market gardeners did not follow the practice used in some more populous nations of fertilising with human faecal material, zapping food with high-pressure steam wasn't really necessary. The pressure cooker took fresh, green peas, beans, broccoli and so forth, removed the crispness that was so unacceptable in traditional 'British' cuisine, and turned them into something like a khaki-green semi sludge. Meat would take on a sort of ultimately dead, grey hue.

My mother, whose passions were playing the piano and growing roses, was not an enthusiastic cook but she did experiment extensively with her pressure cooker. Pressure-cooked curried sausages have to be at the lower end of the human gustatory experience. The gadget was, however, useful as a portable steriliser for surgical equipment and glassware, and I used pressure cookers extensively in my early days as a veterinary scientist and microbiologist.

Now we cook food briefly by using the low-pressure steam that comes directly off water heated in a microwave or on a stovetop, in such a way as to allow the retention of flavour, consistency and colour. The dissemination of cuisines resulting from both immigration and travel means that many homes will boast a bamboo steamer that is, in particular, used to introduce the heat and taste of root ginger into a variety of dishes. Also on the kitchen bench we are likely to find a domestic espresso machine with an attachment for steaming milk. Mishandling this gadget gives many a reminder that steam can scald. On an inexpensive version we owned, I allowed the water level to get too low, with disastrous consequences that led to distorted metal and a smell of burnt rubber. Another common source of minor scalds is the ubiquitous steam iron.

As an adolescent schoolboy gangling on the platform near the equally impatient Thomas the Tank engine, I was always conscious that hissing, high-pressure steam is a powerful and dangerous genie. There's a very human sense of being in the presence of something live as the engineer lets off excess steam or the safety valve activates, a reminder that, though we may have the illusion of being in control, the possibility that a critical component could fail disastrously or that we might find ourselves in the wrong place at the wrong time adds an element of danger and unpredictability.

I don't know if Native Americans ever did call steam locomotives 'iron horses' outside the dialogue written for western movies, but it's not a bad analogy. Even so, the characteristic steam engine *choo-choo* sound seems gentle in a contemporary world battered by violent assaults on the ears and other senses. To some extent it recalls the whooshing noises we all experience in the womb. When I was very young, the city of Brisbane still operated some genuine steamrollers, perhaps because of the oil shortages during World War II. Even now, I can recall their accelerating *chuff-chuff-chuff* with pleasure.

Though steam engines are just machines—and, by modern standards, primitive machines at that—they seem just a little more human than much of the technology that dominates our highly mechanised and electronic age. While there may be no good economic reason to do so, many have devoted their spare time to restoring steam locomotives of various types and to keeping at least some segments of rural railways open so that even the youngest child can have the experience of riding behind a smoke-belching choo-choo. Thomas the Tank Engine doesn't look to be in any danger of being replaced by Denny the Dinky Diesel in the childhood imagination. Kids love to ride *Puffing Billy* on the 762-millimetre narrow gauge railway in the Dandenong Ranges east of Melbourne, or to take the train to Paradise pulled by the Strasburg Railroad's Baldwin locomotive in Pennsylvania's Amish country. At the Franklin Institute, the Baldwin 60 000 still moves down the tracks. Children of all ages cram into the massive cabin, a recording plays hissing and chuffing noises, the whistle sounds, and an electric motor moves her about three metres forward, then back again. Everyone is delighted.

While steamrollers, steamships and steam locomotives belong to history, steam remains a very important intermediary in contemporary life. Steam catapults propel navy fighters into the air from the shifting decks of aircraft carriers. The cappuccino/latte and fresh green vegetable sets would be lost without steam. The heat from burning coal or (particularly in France and Japan) nuclear fission produces the steam to power turbine-driven electric generators. Sunlight focused on water through a lens can provide sufficient steam to drive a small turbine. Travelling light, a hand-held steamer is very useful for de-wrinkling clothing. Steam is used to clean everything from household drapes and carpets to historic buildings.

Steam micro-machines that work by vaporising a tiny amount of water have a potential use as 'nano' on/off switches. Contemplating contemporary politics can induce the metaphorical experience of causing 'steam to come out our ears'.

Australia's iconic ballad *Waltzing Matilda* has its swagman anti-hero waiting 'while the billy boiled'—and produced steam. Like the coffee pot on the campfire in a John Wayne western, the 'billy' is a can with a wire handle and a lid that, filled with water and hung over burning wood, is used to indulge the more 'British' passion for tea drinking. Perhaps one of the better omens for the future is that more and more US restaurants now recognise that the optimal way to make that infusion known as a 'good cup of tea' is to pour steaming hot water over the leaves.

Hearth and Home

On chip heaters and auld reekie

A small, perfect pleasure is to relax with the Sunday newspaper and a cup of coffee in a well-heated room overlooking (through double-insulated windows) a sleety landscape or a snow-covered garden. Memphis, Tennessee, is as hot and humid in summer as the sub-tropical, coastal city of Brisbane, Queensland, where I grew up. The big difference is that Memphis has an inland, continental climate and the winters can be cold. On the eve of our first Tennessee Christmas the pipes froze, then burst, and we had to call a plumber. Remarkably, he came.

Our substantial stone-and-stucco Memphis house had been heated through its 100 years by massive hot-water radiators fed by simple convection from an ancient, 'jolly green giant' of a basement furnace. In early days this monster probably burned gas from a local, long-defunct gasworks, but by the time we became its keepers it fed on piped natural gas.

Some of the big, square, low-aspect radiators were placed under long windows, so that you could sit on them, sip your coffee, look out to the garden and feel the warmth come up through the seat of your pants. Because I was raised in a hot climate, there is still something magic for me about cold, snowy days, so long as you don't have to go out and contend with icy roads and freezing winds. When you put all the elements together—good central heating, double-glazed windows and snowscapes—you also realise that this felicitous aesthetic is one that few humans who lived before the modern historical era would have experienced.

My childhood home had no insulation worth mentioning, a corrugated-iron roof, outer walls of over-lapped hardwood planks and, instead of plaster, a lining of tightly fitted tongue and groove boards. Like many of these old 'Queenslanders' (as they're now called), this 1920s house was dressed initially with large verandas that had later been enclosed and it stood more than three metres off the ground on stilts, or stumps. That very practical but rather impermanent construction accommodated to the hilly nature of the cityscape (houses could be five metres off the ground at one end and street level at the other), facilitated air circulation and also kept various creepy-crawlies from accessing the living areas. The open 'under the house' was often the coolest place to be in the Brisbane mid-summer, though you could find yourself sharing the space with spiders, frogs and the occasional large, very poisonous brown snake.

The living and sleeping areas of these wooden houses were hot in summer and, because our sense of temperature is relative to what we normally experience, often seemed cold in winter. People wore light clothing all year round. The 'fashionable' male attire of short shorts and long white socks is a 1960s Australian stereotype, eliciting the same sense of vague sartorial horror as the infamous polyester 'leisure suit' of the Reagan era, but it was much more appropriate for the climate than the two- and three-piece business attire worn by earlier generations of Brisbane office workers.

Public buildings were unheated, though a very few were air-conditioned. Like many other Brisbane houses of that period, the closest thing to a family hearth was a large, black, cast-iron wood-burning cooking stove sited in a vented, galvanised-iron alcove attached to the back wall of the kitchen. The intent of this primitive 'stove capsule' was to minimise fire risk, as wooden houses are highly flammable. In earlier 'Queenslanders', the kitchen was often detached, so that it could burn down without consuming the living areas. Another heat generator located downstairs was the wood-fired copper cauldron, or clothes boiler, that was lit weekly: on washing day. Along with the 'chip-heater' in the bathroom that burned leaves, twigs and wood chips to provide hot water for a shower, the wood stove and the boiler gave way to electrical appliances in the 1950s, the era that marked the beginning of the liberation of Australian women from endless domestic drudgery. Otherwise, the only heating needed in even the depths of winter was supplied by a single-bar electric radiator or two, or by small electric fan heaters. Mosquito nets in summer were more important than doonas in winter!

In the mid-1960s, we travelled from sub-tropical Brisbane to cool-temperate Edinburgh. We rented the converted top floor of a three-storey plus basement sandstone Victorian at 1 South Learmonth Gardens, right at the western edge of the magnificent Georgian New Town that some will know from Sandy McCall Smith's *Espresso Tales* and *44 Scotland Street*, an address on the other side of New Town. Our front bay windows looked down a long, steep hill to the extraordinary Scot's Baronial pile of Fettes College where the long-serving British prime minister Tony Blair must have then been in residence as a boarder, I realised when I checked his biography. If he was taking cold showers at that formative stage of his life, it's a pity he didn't repeat the experience before buying into the disastrous military adventure in Iraq. Along from Fettes is the village of Stockbridge and the Water of Leith that flows on into the Forth of Firth at the village of Cramond.

Some of our more established friends lived in modern Wimpey-built tract houses on the outskirts of town. They had central heating, but the Britain experienced by transient adventurers like us wasn't characterised by warm interiors in those distant 1960s days. The heating system for our top-floor flat consisted of single bar radiators and a fan heater, the same technology that we had in Brisbane. The difference was that ice formed on the inside of the poorly fitting, very original windows in winter. We wore heavy tweeds and woollens and slept under goose feathers to maintain our body temperature. Before we brought our first child home from Edinburgh's Western General Hospital on a cold January day, the district nurse, following standard procedures, visited to ensure that there was adequate warmth in at least one room for the new baby. By then we had added an electric convection heater that remained switched on through the winter days and nights.

Our flat still had one open fireplace, though it no longer functioned because the flue had been blocked off. As with most northern European cities, Edinburgh interiors were heated by coal fires through the eighteenth and nineteenth and the first half of the twentieth centuries. The other name for Edinburgh was 'Auld Reekie' (old smokey). A traditional Scottish blessing is 'Lang may your lum reek'—'long may your chimney smoke'. Engravings and photographs from those times show spiralling plumes of black smoke rising from the mass of chimney pots that still grace the inner city buildings. When that dense smoke combined in winter with the 'har'—the sea mist that rolls in regularly from the Firth of Forth—Edinburgh must have been both 'gie dreich' (very bleak) and bad for anyone with respiratory problems.

In the end, though, the air pollution associated with burning coal in open grates became unacceptable. The great London smog of 1952 that is thought to have resulted in the death of some 4000 people marked the closure of that era. By the 1960s, most of Edinburgh's smoking chimney pots were history. The only substantial change in air quality that we noticed from time to time in South Learmonth Gardens was a pervasive, yeasty smell coming from one or other of the many breweries. Once, we looked out our windows to the surprising sight of a fully kitted-out municipal firefighter perched precariously on the sloping roof of a tall building diagonally opposite, monitoring the aftermath of a no-doubt terrifying conflagration in a little used, unswept, domestic chimney.

Coming from Australia, which has none of that type of history, we spent many of our weekends climbing over ruined castles and mediaeval tower houses in the Scottish country-side. The great halls of those residences had large, open fireplaces that were obviously designed for burning big logs. Much of that timber cover had been consumed, and was no longer apparent in rural, twentieth-century Scotland. Having just one large, heated hall in a fortified tower house would certainly have encouraged togetherness, a habit that probably contributed to mutual survival in that brutal and treacherous era of Scotland's history. Walking through more intact historic residences that had housed the well-to-do, we realised that a big advantage of the canopied beds the aristocrats shared would have been to conserve body heat.

As both rural and city houses became more like those we live in today, having small fireplaces in different rooms allowed both privacy and the 'personal space' that is necessary if you are to have the time and peace to think. It's no accident that the elegant, urban architecture of the New Town, built from 1776 to 1840, coincided with the intellectual flowering that we know as the Edinburgh enlightenment. The names of the philosopher David Hume and the economist Adam Smith will be familiar to most, and there were many others.

Life for the eighteenth-century poor, though, could still be both cold and hard, particularly in the 'closes' of the High Street and industrial areas like the Grassmarket, the Haymarket and the Cowgate that are located on the low side of the old town that stretches down from Edinburgh Castle, past the great Cathedral church of St Giles with its imposing, open steeple crown, to the Royal Palace of Holyrood, where Mary Queen of Scots had her first husband, Lord Darnley, murdered. We all know of Catholic Mary's fate at the decree of her Protestant cousin, Elizabeth I of England. Mary's other nemesis, the uncompromising Calvinist John Knox, is evidently interred somewhere under the St Giles car park! Though we were not aware of the fact when we lived in Scotland, later genealogy research told us that the tailor Henry Edwards, Penny's great, great grandfather, emigrated (in 1852) from the Cowgate to seek a better life in Australia. Some of these old neighbourhoods, which will also be familiar to readers of Ian Rankin's gritty Inspector Rebus detective novels, are intact and the buildings still look grim, at least from the outside.

My very able (slightly younger) scientific collaborator, Hugh Reid, a son and grandson of former Moderators of the Church of Scotland (the head of the politically powerful Kirk), lived with an attractive young lady in an eighteenth-century 'working-man's' residence. As I recall, their two rooms in a smoke-stained five-storey tenement shared a single toilet and water source with others who lived on the same stair. That space occupied by a couple of students would formerly have housed a whole family, with at least some of the heat that they required to see them through winter coming from their own bodies. As in *Angela's Ashes*, poor children huddled together in the same bed. In other, modest houses, a child or a single person would often sleep in a curtained-off, shelf-like alcove in the kitchen. Part of a snooty Edinburgh cultural myth was that some Glasgow kids were still being sewn into their long underwear for the winter! True or not, such tales of the 'great unwashed' had some credence with expatriate colonials. British bathing habits, and the difficulty of getting a hot bath in many boarding houses and the like, were standing jokes with impecunious Australians.

Travelling south through the Scottish borders, the enthusiastic tourist soon comes upon traces of a much earlier and, at least for some, better-heated culture. Our occasional path from Edinburgh to England and the channel ports took us down the A68, past the ruined Abbey at Jedburgh, across the border into Northumberland, then to the Roman emperor Hadrian's wall and the associated fort at Corbridge, or Costorpitum as it was called in ancient times. Upper-class Romans who were transplanted to such outposts of empire from their temperate Mediterranean homes insisted on a degree of comfort. The heating problem was solved by the hypocaust which, in its more sophisticated form, elevated the mosaic-covered concrete floor on piles to give an airspace below. A furnace that burned brush or wood was then used to heat the 'crawl space' air and warm the room. Simpler versions had the hot air circulating through one or more flues in the walls.

No doubt thinking back to a warmer climate, the Romans also enjoyed bathing. The technology used in the smaller bath-houses attached to regional Roman villas and forts used the furnace/hypocaust (or flue) approach to provide an experience that was evidently equivalent to that of a modern Turkish bath. The city of Bath, which shares the Georgian facades of Edinburgh's New Town, was built around Roman baths fed by hot springs, as was the case in many of what were later to become the fashionable spa towns of Europe.

Though the Roman officers and their families enjoyed under-floor heating in their villas, that was not necessarily the case for the tough, effective common soldiers who were the 'sharp end' of imperial power. Even so, though the lower ranks may have slept in buildings that were unheated, the Romans built public baths that could be accessed by both their own people and by the better-off members of the local population. Life was no doubt much harder and colder for the slaves, who were an essential labour force in such societies but were considered of little account by their masters.

While other approaches are generally used to provide under-floor heating today, blown hot air is a common feature of modern residential premises. Anyone who travels and has, as most scientists do, stayed for a time in relatively inexpensive hotels or motels will have had the experience of an electric unit that blows hot or cold air and makes a noise like a Boeing 747 in flight. The choice is sleep or freeze. More upmarket hotels may have a central air system that is ostensibly controlled by a thermostat in each room. Sometimes that dial is actually connected to the system, while in other cases it does nothing and is evidently there to give the deluded guest some spurious sense of power.

Though many middle- and upper-class people now live as though a controlled, heated or cooled environment is a divine right, this is not the current reality for many of the poor in the world and was never the case for the vast majority in those generations that came before us. Not long back we had the spectacle of children and families freezing to death in the high country of Pakistan, because there was no way of getting adequate food or shelter to them after the earthquakes that had destroyed the roads, their homes and villages.

Human agencies can also be involved in the deliberate triggering of crises that cause others to freeze. A prime example is the recent confrontation between Russia and its former satellites over natural gas supplies. A few years back we had Texans threatening to do much the same thing so that they could force higher gas prices on those living in the cold, northern parts of the United States. The rhetoric at the time gave the impression that the Texans had no sense of common destiny or shared experience with their fellow US citizens. That was undoubtedly true for the criminals at Enron, who manipulated the availability of electricity to, particularly, Californian residents. Extraordinary arrogance and personal greed were at the core of their beings.

Even those in the prosperous west can be affected by natural disasters, as we saw with Hurricane Katrina. Most long-term Memphis residents have had the experience of losing all electric power and freezing in their homes for a week or two at a time because ice storms bring down cables that are strung between poles. The Memphis Light Gas and Water utility company hasn't buried the power lines because of the high water table. Ice storms happen when rain falling from higher up freezes instantly when it hits surfaces that are much colder because of an air temperature differential at ground level. As a consequence, layers of ice gradually accumulate on the twigs and branches of the trees that are such a great feature of the Memphis landscape.

The ice storm effect is extraordinarily beautiful, at least when the sun comes out, but the problem is that the added weight causes large branches to snap off and fall. Furthermore, because much of the city is built on Mississippi river mudflat, many of the magnificent oaks in the older parts of town are not deeply rooted and can come crashing down. We lay in bed one winter evening listening to the whoosh then thump of falling branches, punctuated by the occasional loud bang as an electricity transformer exploded. A tentative, icy walk the next morning took us down neighbourhood streets blocked by large trees that had been either completely uprooted or snapped off at the base of the trunk. We saw crushed cars, and occasional Federation-era (Victorian/ Edwardian equivalent) 'Arts and Crafts' bungalows or two- to three-storey 'Four Squares' with smashed-in roofs, attics and upper bedrooms. Fortunately nobody died in our early twentieth-century subdivision, but a car-commuter was killed by a falling tree in a newer and more upmarket suburb.

The loss of electricity soon made the house very cold but we were fortunate that, though our green giant of a gas furnace was not functioning because the thermostat operated from mains power, we could have a hot shower as the gas-fired hot water boiler was turned on and off by a simple bi-metal switch that required no electricity. Also, largely for the pleasure of being able to sit in front of a burning log, the previous owners had installed a slow-combustion stove in the modern family room that had been added to the back of the original 1903 house. We hadn't been using what we called the 'black monster' because lighting it on one occasion had set off a recently installed smoke alarm that automatically called the local fire brigade. Having a ten-metre ladder truck with full crew pull up outside our door had seemed a waste of their valuable time, though they said they were bored and needed a run and were quite nice about it. We donated every year thereafter to the firefighters' retirement fund.

That type of simple wood combustion technology constituted our main home heating for the two separate intervals that we spent in Canberra when I was working at the Australian National University. Though they are no doubt fine for occasional 'recreational' purposes, those wood burning stoves are now as anachronistic for use in large cities as the coal fires of 'Auld Reekie'. Which points to a problem. By removing sources of household combustion associated with other than the relatively clean, natural gas from our urban landscapes, we lose our 'heating independence' and make ourselves vulnerable to direct exploitation by the less scrupulous players in the 'big energy' equation. We are also exposed to the direct consequences of war, terrorism and natural disasters like earthquakes (a very real possibility for Memphis) that disrupt piped-in electricity and/or gas supplies.

One possibility that seems particularly appropriate for both Australians and those living in the sunnier parts of the United States is to turn to the local, household-based technology of solar heating. There is clearly a way to go with the development of solar but, even now, a major reason that it is not in much wider use is because non-renewable oil and natural gas supplies are still much cheaper to access. At some stage, that situation will reverse, perhaps very rapidly.

We could start making the switch tomorrow by levying appropriate carbon taxes and by applying the resultant revenues to encourage individual home owners to 'go solar', if that happens to be an appropriate option for them. Such taxes could also be used to provide incentives for building more energy-efficient houses, to retro-fit existing living spaces so that they conserve heat in winter and exclude it in summer, and to promote research and development across the whole spectrum of the renewable energy sector. At least, while current climate conditions apply, though, solar is unlikely to be a substantial part of the solution for grey, misty cities like Edinburgh. A standing Edinburgh joke is: 'A great summer last year, and it was on a Wednesday!'

The Iceman Cometh

On Monticello and Silent Knights

Being a recreational theatre-goer
rather than a literary scholar, I had somehow come to the
idea that the 'iceman' in the title of Eugene O'Neill's
powerful play was death—either physical death or the death
of the human spirit. It seems, though, that O'Neill deliber-
ately left the definition ambiguous in order to convey shades
of meaning. *The Iceman Cometh* was written in 1939, towards
the end of the era when well set-up sons of toil still delivered
heavy blocks of ice to household doorsteps and even into
domestic kitchens. Urban legend has it that, along with the
milkman, the icemen served society in ways that have more
to do with the creation of life than with the ending of it. The
song title 'The Frigidaire Can Never Replace the Iceman'
indicates that speculation about the iceman's other possible
role was standard music hall fare in the latter days of vaude-
ville. Among other aspects, *The Iceman Cometh* deals with the

sexual tensions between men and women that can, particularly in the case of the central character Hickey, be massively destructive.

Using modern DNA technology, we could probe whether any *Desperate Housewives* scenario has relevance to the way icemen really lived. Would retrospective analysis provide surprising and discomfiting insights into the identities of the alpha males in late nineteenth- and early twentieth-century society? Such a study might be tinged with a certain sense of sadness, especially if we could indeed identify the genetic profiles of ancestral icemen who maintained a high level of activity in the decades immediately before, then after, the years 1914 to 1918 that so depleted the numbers of able-bodied young men.

Back in the first half of the twentieth century, the ice chest that took a large block of ice as the sole occupant of its top compartment was the main means of keeping perishables cool and fresh in most kitchens. It was very important to avoid a flood in the kitchen by emptying the container holding the melt water twice daily. Those varnished oak, metal-lined cabinets can still be found in antique stores, along with the substantial ice tongs that allowed the ten to twenty kilogram blocks to be carried on a hessian-covered shoulder from the horse-drawn wagon or motor truck. The ice picks that were used both for sizing the blocks to the ice chest and for chipping off bits to chill a cocktail also survive in large numbers. Strongly made, wood-handled steel stilettos, they were the weapon of choice for some mafia hitmen and political assassins, though Leon Trotsky was killed with an ice axe, not an ice pick. The psychiatrist Walter Freeman, who operated with disastrous consequences on President Kennedy's younger sister, Rosemary, is said to have performed prefrontal lobotomies with an ice pick hammered into the brain through the tear duct of the eye. He later designed a more elegant lobotomy instrument that was presumably easier to sterilise.

Even the domestic ice chest, which seems so primitive to us now, was a very recent innovation in the tide of human affairs. Large eighteenth-century houses in areas with cold winter climates like Scotland or New England often had a substantial 'ice house'. Generally oval or circular, lined with stone and built around a central floor drain, the ice house was separated from the main dwelling and dug into a natural, or artificially created insulating earth bank. Filled with ice and snow in winter, it acted as a cool chamber through at least most of the summer. Wealthy landowners in the warmer south of England paid substantial amounts for block ice that was cut and shipped from Scandinavia. Many of these humble, utilitarian structures outlive the elegant manor houses and the mansions that they served. Thomas Jefferson's Monticello survives, along with an ice house that is five metres deep by 2.5 metres across and took sixty-two wagon-loads of river ice. Apart from the effort involved, filling the Monticello ice house annually from natural sources in northern Virginia could be an uncertain prospect now.

Lacking any such possibility of access to natural ice, AP McCormick, working on the Western Australian gold-fields, invented the Coolgardie safe. This consisted of a perforated metal or mesh cabinet surrounded by wet hessian that had one end in a pan of water. When it was placed in a breezy spot, the process of evaporation cooled the contents. My grandparents had a variant of the Coolgardie safe hanging from the beams under their house, and I expect it was the only means they had to keep meat and milk fresh for even a short time when they first moved in to their early twentieth-century 'worker's cottage'. These 'meat safes' were either suspended or placed on legs standing in ant cups to stop those ubiquitous scavengers getting to the contents.

Because of the problem of bacterial putrefaction, meat was traditionally smoked, preserved in brine, pumped with sodium nitrite, or dried in the sun in strips (pemmican). Products like bacon, ham, corned beef, rollmops, pickled herring, smoked haddock, lutefisk, pickled onions, gherkins, pickled eggs and so forth continue to add a certain taste to life—although, if taken in constant and substantial excess, they can be implicated in a variety of medical conditions from high blood pressure to stomach cancer. Still, anyone who has been raised on such a diet may have real nostalgia for the occasional plate of corned beef and carrots, or a breakfast kipper. Those particular food cravings can be hard to understand for some who have lived only in contemporary times. More familiar to most are the curries and hot sauces that, designed for a similar purpose, are now well established and add colour and piquancy to the cuisines of most countries.

The limited storage capacity of the ice chest, and the uncertainty that the ice would last till the next delivery, meant that people bought perishables in small quantities. The mother of the family, or the maid, would make daily trips to the local butcher, fish market and green grocer. Milk, like ice, was generally delivered, often in a horse drawn, two-wheeled tank filled at a local dairy. The milkman operated the tap used to fill a jug, or 'billycan'. Suburbs were dotted with a variety of small stores that were in easy walking distance. These family businesses were, of course, quickly killed off by the automobile, refrigeration and the supermarket. Many of the Victorian and Edwardian store fronts and their associated metal awnings survive in heritage-protected areas of central Melbourne, though they are now often used as dwellings. Local bakeries, and in-city dairies where cows were milked daily, have long been converted for residential or business purposes.

The mechanical parts of an ice-making plant, a household refrigerator or an air-conditioner allow a liquid refrigerant to vaporise and condense in a continuous cycle. Conversion to the gaseous phase involves the uptake of heat to give the cooling effect, while liquefaction in modern refrigeration units requires the input of energy and heat production from a compressor driven by an electric motor. A contemporary air-conditioner has the compressor and heat-dissipating fan outside the house, while the cooling coil and the circulation fan are inside and feed directly into the room or ducting system.

Historically older 'intermittent absorption' refrigeration plants used a cycle involving the varied application of heat from an electric or kerosene burner to an ammonia/water solution confined in a closed system. The cooling results from successive cycles of liquefaction to give anhydrous ammonia, followed by evaporation and the return of the gas to the ammonia/water mix. The changes in pressure associated with the provision or absence of heat, together with the expansion and contraction from fluid to gas then back to fluid, are a primary force driving the reaction. While this mechanism gives very effective refrigeration, it's also inefficient and expensive to run in energy terms.

Even so, because the basic units were cheap and easy to manufacture, intermittent absorption powered the first domestic refrigerator, the 'Silent Knight', that appeared in many Australian households during the 1940s and 50s. The lack of a motor and compressor meant that it made no sound while running. Burning kerosene as an energy source also allowed Silent Knights to be used in regions that lacked a high voltage electricity supply, where a few battered relics continue to soldier on.

The nature of refrigerants has changed over the years. Some of the earliest systems used ether, which was potentially explosive. Anhydrous ammonia boils off at minus 33 degrees Celsius, plenty cold enough to freeze a side of beef or lamb. In ice works this was used to refrigerate a coolant such as saline, which has a low freezing point, for circulation around block-sized moulds filled with drinking-quality water. After the expenditure of more energy by the horse drawing the wagon, the iceman with his tongs and the eventual household recipient wielding an ice pick a little of the product could finally end up in someone's gin and tonic. All that urban labour and muscular effort has been replaced by remote, largely automated power plants and networks of electricity lines that transfer the energy from spinning, steam turbine-driven generators to the compressor and ice-maker of a modern refrigerator.

The chlorofluorocarbon Freon, which found subsequent widespread use in cooling systems, had to be abandoned because of the effect the CFCs were having on the ozone layer of the stratosphere. The fact that we were able to mobilise the human family to make the necessary change to less damaging products is one cause for optimism as we face the issue of global warming, though the adjustments that were required on the part of both businesses and individuals were minuscule when we contemplate what will need to be done to stabilise atmospheric CO_2 levels. Interestingly, CO_2 itself is being promoted as a relatively non-polluting refrigerant for, particularly, automobile air-conditioning systems.

The widespread availability of refrigeration changed whole societies. The introduction of refrigerated ships from the early 1880s allowed otherwise unpreserved meat and butter to be transported from distant outposts of empire like Australia and New Zealand to the United Kingdom. This imperial trading block survived until Britain entered the European Economic Community (EEC) in 1973, a move that caused major trauma for, particularly, New Zealand primary producers. The removal of the constraints imposed by those economic and cultural ties led to the abrupt reorientation of export markets and a rapid weakening of the linkages between what my Essex-born Byford grandparents referred to as 'home' and the English-speaking societies of the southern hemisphere. Any Australian who has stood in the non-EEC line at Heathrow airport is acutely aware of this change, though some of our senior politicians who move from aeroplane to VIP lounge and beyond continue to miss the point and delude themselves into thinking we have some meaningful 'special relationship' with British governments.

When supermarkets replaced corner stores, the effect was much more widespread in warmer climates. A Silent Knight was sitting in the corner of my parents' kitchen in sub-tropical Brisbane by the beginning of the 1950s but when Penny and I moved to Edinburgh in 1967, we just put our milk bottle outside on the window ledge at night. Then, after the birth of our two sons, we bought a standard-size (for Scotland) domestic refrigerator, much like the bar fridge found in most hotel rooms. The local fishmongers and butchers (fleshers) displayed their products in simple, un-refrigerated cabinets and had game, like pheasant and hare, 'hanging', a polite euphemism for using bacterial fermenta-tion to soften the meat. Visiting recently, we noticed that much of that old, and to us comfortingly archaic, culture has given way to ubiquitous supermarkets.

The enormous progress made in biomedical research and health care over the past century or so would have been inconceivable without refrigerators and freezers. Cancer research would have gone nowhere. The delivery of live vaccines, like those to prevent poliomyelitis, measles, rubella and yellow fever, is not possible without an established 'cold chain'. This is a major consideration when such control programs are implemented in the poorer, less accessible regions of the planet. We laboratory scientists keep our reagents, sera, monoclonal antibodies, tissue culture media, antibiotic solutions and so forth as liquids at 4 degrees Celsius or frozen at minus 20 or minus 70 to 80 degrees Celsius. Mech-anical freezers use a double compressor system to get down to those low temperatures. These resources are so valuable that the top research institutes, like St Jude Children's Research Hospital, have their key storage units connected to back-up diesel-powered generators. Some of the cancer tissues held there have now been kept frozen for more than forty years, an operation that is not inexpensive but resulted in a recent, major scientific breakthrough that I discussed in my previous book, *The Beginner's Guide to Winning the Nobel Prize*.

In addition, irreplaceable research reagents like genetically engineered mouse embryos, sophisticated virus vectors for experimental therapy, vaccine seed stocks and so forth are stored in liquid nitrogen cabinets and in Dewars (big, vented thermos flasks) at less than 180 degrees Celsius. The same is true for the *in vitro* fertilised human embryos that cause a small minority of devoutly religious people a great deal of moral agony while delighting the majority who see a childless couple embrace the arrival of a previously denied infant. As a young veterinarian, I was involved in collecting horse and bovine sperm for freezing and later use in artificial insemination to improve livestock quality and productivity. The vaporisation of solid CO_2 (dry ice) was used routinely to protect such biological specimens as they were transported from one place, or one country, to another, though many airlines are now reluctant to carry substantial amounts of dry ice, and liquid nitrogen flasks have found much wider application.

Perhaps the biggest global change, though, has been the introduction of air-conditioning, an advance that allowed the lethargic, torpid communities of the hotter parts of the planet to become more like the cooler, generally northern societies.

Growing up in Brisbane of the 1940s, we spent our summers sleeping under mosquito nets with the windows wide open. Even well-to-do neighbourhoods in Sydney's fashionable North Shore suburbs 700 kilometres closer to the Antarctic often had only two bedrooms in those days, one for the parents and the other for the girls in the family, while the boys bunked down in open 'sleep-outs'. In large families, the same could be true for the man of the house, partly because of space and partly as a reflection of the uncertainties associated with then-current methods of contraception.

Government buildings were made of stone or wood, had deep verandas and large windows, the type of construction that can still be seen in Singapore near the cricket ground and in the older parts of many outposts of Empire. As late as the 1960s, the available air-conditioning in Brisbane was largely restricted to butcher's shops, movie theatres and big department stores. Now, when we visit a dynamic, tropical city like Singapore, we spend most of our time living and working in air-conditioned hotels, offices and lecture theatres that are indistinguishable from comparable facilities in Boston or Miami. The old, open, colonial-style State Treasury building in Brisbane has been transformed into an hermetically sealed casino.

The back of our century-old Memphis house had previously been graced by outside sleeping porches, which were screened because the city was nearly wiped out twice in the nineteenth century by the mosquito-borne yellow fever virus. Though some of these rather flimsy structures survived in the neighbourhood and we could trace the outlines of what had been an external door in the plaster-work of the upper level of our house, most had been pulled down or allowed to collapse with the advent of air-conditioning. Also, unlike their Sydney North Shore contemporaries, the homes of the Memphis middle class were bigger and had more bedrooms. A sleep-out was simply too cold in winter for even the hardiest boys.

In the heat of summer, external ceiling fans provided some cooling effect for those lounging languidly in rockers and porch swings on the substantial front verandas of these Federation-era 'Four Squares', 'Queen Anne' and 'California' bungalows. It should surprise nobody who visits in June to September to learn that the famous Hunter fan company was founded in Memphis. A somewhat irreverent theory has it that the iced tea found so ubiquitously throughout the American South served the additional function of allowing porch-sitting, middle-class, church-going ladies to sip slowly on a suitably diluted shot of bourbon whisky (or moonshine) through those long, hot, humid evenings without giving the game away.

Many of the bigger residences also had massive attic fans, with blades more than a metre and a half in diameter, that were used to draw the cool night air into the house through an outer, open-slatted front door. An older friend told us that, before retiring for the night, her father would hose down the plants in front of the house, then turn on the attic fan to give the evaporative cooling effect that was used in the design of the Coolgardie safe. Because of the difficulty in running standard-size air-conditioning ducts in an old house, we had three separate central units, one venting through the floor with the ducts running in the basement, another in the attic and ceiling of the upper storey, and a third serving a recent addition at the back. As Memphis accesses hydroelectric power from the Tennessee Valley Authority and the house was well insulated, these were surprisingly cheap to run.

But there's the rub: air-conditioning and refrigeration take energy. The same is true for the compression of nitrogen or CO_2 to the liquid or solid (dry ice) state. The ultimate cooling effect is more than balanced by the release of heat in the production process. In the absence of the hydroelectric, solar, wind or nuclear power that is as yet available to relatively few people, we have to use coal, oil or natural gas to generate electricity. As an automobile air-conditioner is driven by a belt from the engine, the harder the cooling system works the more the gas mileage will drop. It doesn't matter whether fossil hydrocarbons feed local furnaces that provide steam for a compressor in an ice factory or a ship, or fuel the turbines in an electricity plant, there is the same potential for CO_2 release into the atmosphere. Also, setting aside any bad jokes about the possible sexual athleticism of the now historic urban iceman, the refrigeration-dependant scientific advances in vaccine research and therapeutics that have contributed to improved infant and adult survival are clearly a factor in the massive, four-fold increase in the size of the human family over the past century. The equation linking population size and global warming is a most inconvenient truth that many economists, politicians and religious leaders (among others) do not wish to acknowledge or even discuss.

The desire to stay cool in our houses and workplaces while enjoying good health has been just one of the many fossil fuel-related activities that have contributed to heating the planet. The tension between global warming and the need of homeotherms like us to maintain a body temperature around 37.5 degrees

Celsius creates a vicious cycle that must be addressed with all the means at our disposal. Apart from increasing the pace of efforts to develop renewable energy sources, much more could be done now to design buildings in ways that make greater use of breezes, the cool night air and the more stable temperature of the deep earth itself. We have to be cold, hard devotees of evidence-based reality as we strive to develop solutions to the problem of planetary warming that will allow us to continue living in safe, comfortable human environments.

What role will the iceman play in the human future? Those strong men with their ice tongs belong to a particular, brief period in the history of labour and are, unless a few survive in retirement communities and nursing homes, long gone from our midst. Regular deliveries of block ice to households stopped almost everywhere between fifty and seventy years ago. We buy bags of crushed 'party' ice from the supermarket or petrol station, but they don't require much physical strength to carry and won't do much damage if they happen to slip from the grasp and crash onto a big toe.

The imagery of 'the iceman' as the angel of death might bring to mind the 30 000 or more who died in the 2003 European heat wave. Though we should be a little circumspect when we read such numbers as they refer to excess mortality rates (some may have already been very ill), it will be a great tragedy if future generations look on those deaths as 'just the beginning'. As any desert dweller knows—and all Australians live on the rim of a vast desert—heat can kill, and kill quickly. Unlike Miami, Cairns or even Washington DC, mid to northern European cities are not designed to deal with prolonged, very hot weather.

Paradoxically, as global warming proceeds, we may be in for out-of-season encounters with the iceman of winter. In general, weather patterns are becoming more unpredictable. In 2006 snow fell in November in sub-tropical southern Queensland and on Christmas day in Victoria, in the midst of the southern hemisphere summer. The same increase in ambient temperatures that revealed (in 1991) the 5300-year-old, bronze-age Iceman of Oetzi in a melting Austrian glacier is also diminishing the northern ice packs. The consequent increase in the volume of heavier, fresh water sinking to the depths of the ocean may have the paradoxical effect of stopping the rise and fall of ocean currents that provides the Atlantic conveyer and the warm Gulf Stream. If that circulation stops, Britain could become much colder. Edinburgh and Moscow are at the same latitude. Stay in an older London hotel, or look at the prominent outside plumbing while driving through the picturesque countryside, and it is very obvious that, solely from the aspect of changes in construction and engineering, adjusting to a much colder reality would involve enormous costs.

Like many great plays, Eugene O'Neill's *The Iceman Cometh* deals with deep aspirations and personal disappointments. At times his characters are in denial, at times they recognise their inadequacies and self-deceptions. If the various manifestations of the iceman are to be kept in their respective places we must all strive to put fantasy behind us, confront the reality of what is happening now and embrace the need for rapid technological and behavioural change.

Night Lights

On candles and whalers

My paternal grandmother lived in a little house built just behind the old family home where my father, then my brother and I, grew up. Widowed early with four sons, Helen Doherty (nee Chippendale) was accustomed to a very simple, frugal life. Brisbane has no long summer evenings, and our garden was devoid of the magic fireflies that graced the Philadelphia garden that Penny and I enjoyed many years later. On occasions when she ate dinner with us, Helen lit her short path home with an antique tin lantern. After opening one of the four glass panels, she put a match to the white candle, then set off into the dark sub-tropical night. My vivid memory is of a slow-moving, bent figure holding a small, flickering orange flame.

Any human living over the two or three millennia before the twentieth century would find such an image familiar, as might many in the developing world today. In our contemporary western culture candlelight is associated with incense, churches, romantic dinners, camping and electricity black-outs, rather than with normal, day-to-day practicalities. Religious rituals use the readily extinguished candle to remind us of our transience. We no longer depend on candlelight as a practical reality, and are less aware of the dangers inherent in burning our candle of life at both ends. At the same time, modern medicine has tended to slow that rate of combustion.

Burning candles in sealed bell jars was one of several incisive experimental approaches that allowed the eighteenth-century French scientist Antoine Lavoisier to characterise the gaseous elements of air, particularly oxygen (O_2). The flame burns bright, then snuffs out when the oxygen is consumed. The candle gives light and hot air, though only in some Dickensian *Little Nell* scenario with cold hands cupped around a small flame could this be thought to provide much in the way of heat. What Lavoisier did share in a horribly real way with the imagined world of Charles Dickens was the fate of Sydney Carton in *A Tale of Two Cities*. Coming from the aristocracy, much of Lavoisier's wealth was derived from being a 'tax farmer', or tax collector, for the Crown. The overheated, early days of the French Revolution were not a good time to be in that particular line of work.

A story has it that, in condemning Lavoisier to the cruel embrace of Madame La Guillotine, the court stated, 'The revolution has no need of savants.' Despite that, I couldn't find indications that any of Lavoisier's fellow members of the Sciences section of the Académie Française, an organisation that had then been in existence for more than 120 years, shared his fate as a victim of the Terror. At the age of fifty-one, Lavoisier's light was snuffed out by, among other traumas, his clearly superior brain being instantaneously deprived of its blood supply and thus oxygen!

Light achieved by burning wood, rushes, a candle, oil or gas in oxygen was the only artificial illumination available to the generations of humans that lived in the times before Thomas Edison discovered, in 1879, that passing electricity through a carbon filament in an oxygen-free mixture of inert gases gave a bright, steady light. He followed earlier experiments by Humphrey Davey, Joseph Swan and others who showed that an electric discharge between two carbon (C) electrodes causes them to vaporise and burn. As they burn, the carbon combines with oxygen to give the greenhouse gas CO_2. However, though they found some use for street illumination and theatrical events, the brilliant light emitted by such open carbon arcs was not practicable for daily domestic use. The underlying engineering was complex, the carbon rods were quickly oxidised and consumed, and arc lamps use a lot of electricity.

The eighteenth and nineteenth centuries ushered in the era of adequate home lighting. Starting with municipal gaslights along London's Pall Mall in 1807 and in Baltimore in 1815, coal gas was piped to increasing numbers of well-to-do houses through the first half of the nineteenth century. Commercial coal gas was generated by baking coal in beehive ovens to provide coke and methane. Coke was used to fire the foundries that provided iron and steel, the primary materials of the early mechanical age. That use of coal continued for only a few short years and most of the gas plants were demolished long ago.

Methane (CH_4) is generated naturally in coal seams, and miners were always at risk of methane explosions. Methane is odourless, and the naked flame of a burning candle or acetylene lamp mounted at the front of the miner's helmet added to that danger. The gas acetylene (C_2H_2) is generated by dripping water onto calcium carbide (CaC_2), with the brightness of the light being controlled by the rate of drip. Another gas, the non-flammable, odorless carbon monoxide (CO), was, however, the reason that miners traditionally carried canaries with them to warn of 'bad air'. Now, miners use battery-powered electric lamps.

Our contemporary experience of flickering gaslight is pretty much restricted to butane or propane cigarette lighters (also on the way out as we reject smoking), ornamental gas logs, garden lamps at evening parties, camping trips and watching Ingrid Bergman and Charles Boyer in the classic 1944 movie *Gaslight*. Natural gas (mostly methane, but containing some propane, butane and ethane) is, at least till it runs out, a cheaper and cleaner resource for cooking and heating purposes. Most coal is now used to produce the super-heated steam that drives electricity-generating turbines in facilities remote from the city landscape, but coal-burning plants continue to be a problem as we seek to minimise the production of 'greenhouse' gases.

Through the first half of the Industrial Revolution, the other lighting technology that evolved greatly was that of the oil lamp. Before 1750 or so, candles were a much more desirable form of lighting than the primitive oil lamps of earlier times. The reason was that candles generated less smoke and were less likely to create the bad odours that were associated with burning some of the readily available oils, like castor oil and fish oil. Oil lamps have been around for more than 2000 years. Both bronze (for the well to do) and terracotta examples were found at the Pompeii site that was buried in AD 79 by molten lava from Mount Vesuvius. The ancient oil lamps that we see in museums were open vessels with a handle at one end and, in some cases, an opening at the other. The wicks either floated in the oil or projected through the hole made for that purpose. Wicks were made from all sorts of fibrous material, including the castor plant, which was evidently considered to be an inferior source. Those who were better off would have burned the relatively clean and odourless olive oil but, even then, the wealthy preferred candles.

The candle goes back a long way. The ancient Egyptians dripped beeswax or tallow onto rush stems which, when lit, provided a smoking light. We all understand where beeswax comes from, but what is tallow? I found this recipe on the web: Water is added to cooling, melted beef or mutton fat, the mixture is then kept at a low boil for several hours, and, when it has cooled again, the fat separates into three layers, with tallow being the clear, hard, white substance at the top. Tallow is, in fact, the triglycerides that none of us want circulating at high concentrations in our blood streams. Nothing from animals was wasted in traditional societies and, even now, the only thing that isn't used from the pig in Memphis is the squeal.

Tallow candles were essential for reading, or for socialising at night, in seventeenth-century London. The diary entry of English naval administrator and parliamentarian Samuel Pepys for Monday 16 June 1662, includes: 'I also, with

Mr. Davis, did view my cozen Joyce's tallow, and compared it with the Irish tallow we bought lately, and found ours much more white, but as soft as it; now what is the fault, or whether it be or no a fault, I know not.'

Pepys could have followed the example of his fellow members of the fledgling organization that was to become the Royal Society of London (Britain's National Academy of Science) and made a systematic comparison of the two types of tallow to see which was the better, but he was an observer rather than a experimentalist and it's hard to believe that such an enterprising step would have failed to make it into the diaries. That he and his friend Mr Davis took the trouble to inspect his cousin's tallow speaks to the importance of quality candles for a well-connected, upper middle-class householder in Charles II's Restoration England. Pepys's diaries show him to be a busy man, who used the evening hours to enjoy cards, music, reading and recording the events of the day. None of those activities was possible without light in that era (there were no CD players, only musicians).

The 'fuel base' of most modern candles is no longer tallow but paraffin wax, a heavy, complex, hydrogenated carbon that comes from crude oil. How do candles work? When lit, the wax in the vicinity of the absorbent wick melts and is carried upwards by capillary action. The heat of the flame then causes it to vaporise, and the light comes from this wax vapour burning at the tip of the wick. The white 'smoke' that continues to rise for a little while after the flame is extinguished is vaporised wax.

Oil lamps started to become more important in the latter half of the eighteenth century when it was realised that oil 'harvested' from the sperm whale burned cleanly and gave a bright light that, for instance, allowed an ordinary family to sit around a table and read into the evening hours. Though Inuit tribes and the like had long hunted coastal whales for meat and oil, the Yankee whalers that are so vividly portrayed in *Moby Dick* first came onto the scene in the mid-eighteenth century, when it was realised that there was an increasing demand for whale oil. Unlike fish, whales are warm-blooded, air-breathing mammals that have to maintain a constant body temperature. Because they dive deep to feed in very cold water, they need both the insulation and the energy supply provided by a thick layer of fat.

Anyone who has read *Moby Dick* understands that whaling was a dramatic and dangerous business, though perhaps not as dramatic and dangerous as Gregory Peck's Captain Ahab in the 1956 movie version. The whales were killed using harpoons thrown from open boats that could be towed for miles, and even destroyed, by the dying giant. Whaleboat men were dragged to the ocean floor by the 'sounding' whale if a hand or foot caught in the flying harpoon rope. When the end came, the whale carcass was hauled alongside the mother ship so that the blubber could be cut off in strips (flensing) that were then boiled in large, on-deck, copper cauldrons to give the precious oil. The plume of smoke could be seen a long way off. Men wielded sharp flensing knives and hooks while working off the 'cutting-in stage', a platform suspended from the side of the wooden ship. As all this activity was subject to the vagaries of wind and water characteristic of the open ocean, both fire and injury were constant risks.

Oil wasn't the only product. The keratin baleen of the filter-feeding sperm whales provided the 'whalebone' used in corsets; the oily spermacetti recovered from the head was used for quality candles; and ambergris, a mixture of bile and fat from the gut, was (and still is) an extremely valuable component used in the perfume industry. People didn't bathe as much then, so scent was very important. It's quite a contrast to shift the mind's eye from the pitching deck of a Yankee whaler to the wasp-waisted, tightly corseted, sweetly scented ladies of a well-lit nineteenth-century London or Paris salon. Those elegant social beings would have been appalled if forced to confront the grim realities of deep-water whaling. As with the globalisation of manufacturing in our time, the comfort and elegance of the glitterati depends on the hard and poorly rewarded lives of others.

The New England whalers ranged from Connecticut, Nantucket and Maine to the South Pacific and the cold Antarctic. Climbing over the *Charles W. Morgan* preserved at Mystic Seaport in Connecticut gives a good sense of just how hard these lives actually were. Sailing the open ocean by force of wind alone is a tough business, and the whaler's experience of sharing a bunk and bedding with the man from the next watch is endured today by contemporary racers in the round-the world Volvo classic. That's a modern adventure but, beyond *Moby Dick*, the romance of the nineteenth-century whaling industry that existed principally to provide oil for household lighting continues to enrich our literature and our sense of history. The whalers provisioned in towns like Hobart and Dunedin, leaving, as sailors still do, their mark on those communities. The entertaining and thoughtful

Australian novelist Bryce Courtenay uses this Tasmanian connection as a theme in his *Potato Factory* trilogy. As described in Tom Keneally's *Great Shame*, an American whaler was used in the 1870s to effect the escape of five Fenian revolutionaries who had been transported to Western Australia. More than 130 years later, the south of Ireland is, at least, a republic, though that light is still to shine brightly in Australia.

Deep-water whaling progressively fell by the wayside from the 1850s on. The intense harvesting had dramatically decreased whale numbers, while the demand for oil diminished with the supply of piped coal gas in the inner cities and the more general availability of kerosene, then electricity. Steel stays replaced whalebone in the fashionable corsetry that moulded (mostly) the female form. However, shore-based whaling continued until much later. Ernest Shackleton's heroic 1917 voyage in the tiny *James Caid* ended at the South Georgia whaling station. The story of how all the crew survived the loss of the exploration ship *Endurance* in the Antarctic pack ice is told vividly by the contemporary photographs of the Australian Frank Hurley, in Shackleton's marvellous book *South*, and in several recent movies. Coastal whales were still being processed at Tangalooma in Moreton Bay when I was growing up in the riverside city of Brisbane. Now the nauseating smell of burning blubber and rotting whale meat is gone and Tangalooma is an island tourist resort. Only the Japanese continue large-scale ocean whaling, in an industry that is worth more than 60 million dollars a year, largely because of the demand for whale meat. With some success, organisations like Greenpeace have been doing their best to see that these twenty-first-century whalers are exposed on our TV screens, in the hope that public disdain will bring this activity to an end.

The mid-nineteenth century saw the beginning of our heavy dependence on subterranean oil, initially for lighting and for lubricating the machines of the Industrial Revolution. The kerosene odour from oil lamps and heaters is familiar to most of my generation, and I still catch a whiff of it from time to time at airports. If Jet A fuel isn't identical to kerosene, it certainly smells the same.

Kerosene hurricane lanterns were the emergency lights of my Queensland childhood. So called because the elongated, pear-shaped glass 'chimney' protects the wick from strong winds, these can still be purchased new, with the prices ranging from about US$13 to $70 depending on whether they're made of galvanised metal or the more decorative brass. My retired railwayman grandfather carried a white-painted hurricane lantern, a white pith helmet labelled ARP (air raid protection), and a large rattle to warn the citizenry as he patrolled the local streets during World War II. We were lucky and the bombers never came to Brisbane, although they did hit further north in Townsville and Darwin.

Hurricane lanterns were taken on camping trips, and road works were lit at night by a version with a red glass. The few houses that were still to be electrified in our part of Brisbane during the 1940s used hurricane lanterns or more elaborate oil lamps that survived from the previous century. Like the elderly Miss Marshall who lived across the street from us, some saw no need to abandon the technology they had served them well for seventy or more years. Many kerosene table and wall lamps were very ornate, with polished reflectors and elegant, decorative glasses and metal work. Some are truly beautiful objects that are avidly sought by collectors today.

The shortage of oil caused by World War II meant that a few older people still made their way to the local store in horse-drawn sulkies fitted with kerosene-fuelled buggy lamps. Even then, that seemed archaic. Lights for vehicles evolved during the nineteenth century from candle, then oil-fuelled carriage lamps, to the acetylene/carbide 'bulls eye' lamps that were carried by policemen or mounted on bicycles and the early automobiles. Early cars like the 1908 T-model Fords had acetylene lamps, but Cadillac changed that in 1912 with the introduction of an integrated electric start and lighting system that was soon adopted throughout the industry.

Apart from the switch to kerosene, the nineteenth century also saw a major advance in oil lighting technology with the invention of the mantle, a cylindrical framework of gauze impregnated with thorium and cerium. Contemporary versions are comprised of somewhat different components, but the principle is the same. The heat of the flame causes the mantle to glow brightly, giving a much more satisfactory light. The problem is, though, that the mantles are slowly consumed and, apart from the fact that they have to be replaced fairly regularly, their use in a tent or a closed room means that the occupants are constantly exposed to air carrying the degradation products.

Mantles are still used in pressure lamps like the classic Tilley lantern. A hand pump provides the pressure to give the spray of kerosene or gasoline that is ignited to heat the incandescent mantle. Such oil pressure lamps emit a very bright light, and were the illumination source of choice for early twentieth-century lighthouses. Now, oil is often replaced as a fuel source by butane or propane from a small gas cylinder, a safer and cleaner technology that is used in, for example, the modern Coleman lantern. Even so, there are obvious dangers associated with any naked flame burning in a device that can readily be tipped over or inadvertently moved to a location that is close to flammable material. Kerosene lamps and heaters are still a major cause of house fires and deaths in poorer communities.

The need for fire and flame to enjoy the simple pleasures of reading, writing and conversing through the dark evening hours is now limited to the urban poor and village societies in parts of the developing world, and to those occasions when we seek some form of back-woods or wilderness experience. Though we undoubtedly lost a certain romance and intimacy with the bright illumination provided by electricity, we gained greatly in both convenience and the privilege of breathing much cleaner household air. We broke from the constraint of living from dusk to dawn that has been the reality for most human beings throughout history.

We take it for granted that we can flick a switch and read at any time we choose, a privilege that was available only to the better off among our eighteenth- and nineteenth-century predecessors. Their insights still endure in great novels, in important historical and philosophical works and in enduring statements about the rights of man. Despite the enormous power of the visual media, the contemplation and intellectual activity that are associated with reading and writing are still central to the continued health of human society. When we babysit our grandchildren, it is a true delight to see that they prefer to end their day reading in bed rather than sitting passively in front of a TV screen. The human condition has undoubtedly improved. Just think what it would be like to read in bed by the light of a smoking lamp that burns fish oil.

Imagining the Red Baron

From Biggles to Birdsong

Civilisation went up in flames twice during the twentieth century. Actions taken during and after the conflagrations of 1914–18 and 1939–45 still have consequences for the global political and economic landscapes. War scarred the childhood of succeeding generations. London burned and the aerial Battle of Britain raged during my first year of life. A generation before, my father was almost two when, in April 1915, the Australian Imperial Force landed under fire to confront the Turks at Gallipoli. That May, a 23-year-old Prussian aristocrat who would soon blaze his way across the skies and into the headlines as the Red Baron transferred from the Uhlan cavalry to Kaiser Wilhelm's nascent air force. Military imagery, tales of heroes, the grinding reality of massive human slaughter, and the fact that many adults were grieving, or deeply traumatised, informed the earliest experiences of millions and (looking back) the way that those of us who were very young would later approach life.

In November 1914 a small Australian force took the German colony of North Eastern New Guinea and associated islands (Kaiser-Wilhelmsland) and, that same month, a confrontation between light cruisers saw the first HMAS *Sydney* drive the burning SMS *Emden* ashore on North Keeling Island in the Cocos group.

But World War II came much closer to home. Closing some terrible circle of Mars in November 1941, the second HMAS *Sydney* destroyed the auxiliary cruiser *Kormoran* just off the south of Western Australia. Most of the sailors on the *Kormoran* were rescued while the *Sydney*, last seen still steaming but on fire and down at the bows, disappeared with all hands.

The Japanese attack on Pearl Harbor in December 1941 brought the United States into the war against the Axis powers, with a vengeance. In February 1942, the 'impregnable' British imperial fortress of Singapore fell to overwhelming air superiority and ragged veterans on bicycles who easily negotiated the supposedly impenetrable Malayan jungle. Moving south, US and Australian warships turned back a substantial invasion force headed for Port Moresby at the Battle of the Coral Sea in May. Failure would have put northern Australia in easy reach of land-based bombers. Allied naval–air superiority in the South Pacific was assured in June when Admiral Spruance's US taskforce destroyed four enemy aircraft carriers at Midway. The Japanese tried, in July, to reach Port Moresby by landing on the northern coast of Papua. Australian troops (including my uncle Jack

Byford) slugged it out with them on the Kokoda Trail that crossed the rugged Owen Stanley ranges. Backed up by fresh Australian and American divisions, the Japanese advance was halted and Port Moresby remained in Allied hands.

Unlike many of our European contemporaries, most Australian children never suffered the reality of an aerial attack. Bombs fell on Darwin and on Townsville some 960 kilometres to the north, but Brisbane escaped. Our back garden had an air-raid shelter, dug into an embankment by my father, that was full of spiders and too scary to play in. The closest the physical war came was when the hospital ship *Centaur* was torpedoed and burned (with a loss of 268 lives) off the Cape Moreton Light that marks the seaward bounds of the wide bay where the Brisbane river discharges.

My Byford grandparents' house was in direct line of sight to the skies above Archerfield, Brisbane's original airport. On fine Sunday afternoons in the 1940s, we watched the droning, open-cockpit De Havilland Tiger Moth biplanes doing training circuits and bumps in the not-too-distant airspace. A loud roar and the Mustang fighters of the Citizen's Air Force (equivalent to Air National Guard) City of Brisbane Squadron would flash overhead. Powered by the V12 Rolls-Royce engines that also drove the Lancaster bombers and the Battle of Britain Spitfires and Hurricanes, the North American Mustang could reach the enormous speed of 708 kilometres per hour. Now, a Boeing 747 Jumbo Jet cruises at over 885 kilometres per hour.

On the radio, we listened to the *Air Adventures of Biggles*, the fictional flying ace. A friend whose father worked at Archerfield had a whole set of black moulded models of the different allied and Japanese aircraft that were used to train flight crew, spotters and anti-aircraft gunners. I carved a Mosquito bomber out of balsa wood. We made rubber band-powered propeller planes using razor blades, balsa wood and tissue paper. The solvents in the glue and the liquid dope that hardened and tightened the paper skin no doubt gave us a high, though I don't remember anyone becoming addicted. These flimsy constructions flew, but—unless they were just treated as works of art and hung from the ceiling—they inevitably ended up as a broken mess of torn struts and paper. Still, the combination of concentrated hard work and a high probability of catastrophic failure was good training for a future experimental biologist.

In the air war of my father's infancy, lethal fighting machines like the Red Baron's Fokker Dreidekker (triplane) were also made of wood, fabric and dope. They too could finish as splintered wreckage, with fire and bullet holes adding a grim reality that we did not copy in our models. The motors were much heavier and more durable than rubber bands. Still, the 'Oberusel Rotary' engine from the Red Baron's 'Tripe' that rests at the back of a crowded display case in the basement of London's Imperial War Museum seems incredibly small when we compare it with, for instance, the V12 Rolls-Royce Merlin of twenty years later.

The family bookcase had an odd collection that largely consisted of inherited volumes, technical treatises and presents for long-gone birthdays and Christmases. A magnificent coffee-table book reproduced some of the fine paintings made by World War I Australian war artists. One plate showed *Emden beached and done for*; another the Australian submarine

AE2 that, captained by the gallant Irish Royal Navy officer Dacre Stoker, penetrated the narrow straits of the Bosphorous to enter the Sea of Marmara. Stoker, a distant relative of Bram Stoker of Count Dracula fame, survived the sinking of the AE2 and subsequent imprisonment by the Turks to become a London West End actor. A dramatic painting captures the Australian artillery in action near Ypres. The flyers McNamara and Rutherford are shown lifting their damaged, two-seater biplane off the ground just ahead of the pursuing Turkish cavalry. The originals of these paintings are at Canberra's outstanding war museum, the Australian War Memorial.

Old, bound copies of the *Boys' Own Annual* had stories about aeronauts and serials like 'The Sword of Tolliver Trueblade'. One tattered novel for adolescents described a pre-World War I racing monoplane powered by a mysterious and revolutionary fuel, methanol. The most unlikely of all were two volumes on Germany's World War I fighter pilots, *German War Birds* (by 'Vigilant') and *The Red Knight of Germany* which was about The Red Baron, Manfred von Richthofen. *German War Birds* was a 21st-birthday present to my father from his boyhood friend Stuart Simpson. Like my uncle Charles Byford, Stuart was a prisoner of war after the fall of Singapore but, unlike Charlie, he survived to marry and have his own family.

Despite the hideous slaughter of World War I, both books were written by English authors and portrayed the German flyers as real, believable and courageous human beings. Though millions died anonymous deaths in the trenches, there was still a certain chivalry and panache to the duels overhead between brightly coloured, fragile flying machines. Perhaps that emotional distance between the perceptions of the war in the air and on the ground made the later (1930) publication of these stories acceptable, and even welcome, as people tried to make some sense of the disaster that had ruined so many lives. Another paperback on our shelves was *Foch, the Man of Orleans*, about the French commander of Allied forces on the western front. I doubt that there were any sympathetic British accounts of German Generals Hindenburg and Ludendorf.

German War Birds told the stories of pilots like Max Immelman, who invented the reverse-turn named for him so that he could gain height then drop behind an enemy aircraft and shoot it from the sky; and of Lieutenant Bormann, who developed a passion for knocking down the manned balloons used for artillery spotting. More fortunate than the World War I fighter pilots, the balloon observers were given parachutes so that they could float to the ground. Bormann nearly came to grief when the incendiary bullets fired from his Roland scout ignited a balloon that was booby-trapped with high explosives. The 'Eagle of the Aegean', Lieutenant Eschwege, died when he fell for the same trick. Relaxing in his deck chair and reading a book, Captain Hermann Goering leapt into action to take his Mercedes-engined Halberstadt into combat against a flight of slow two-seaters. After downing an enemy aircraft in flames, he was back reading his book within the hour.

Much of one chapter is devoted to Max Boelcke, who formed the first fighter groups and convinced the Dutch engineer Anton Fokker of the need for a forward-firing machine gun. The 'interruptor gear' that Fokker designed to synchronise the engine's revolutions and the gun's firing mechanism prevented the bullets from hitting the spinning propeller. That gave the Germans aerial superiority for a time, as the pilot could simply aim by pointing the aircraft at his quarry. As a counter, the allies first developed a 'pusher' plane (the De Havilland DH2) that had the engine and propeller behind the pilot and the machine guns in front, then worked out how to replicate Fokker's 'interruptor' strategy for use in more conventional aircraft. I know almost nothing about the World War I allied aces, like Billy Bishop, Albert Ball, Robert Little or Eddie Rickenbacker, but still have my father's books and remember these stories of German war birds.

The most famous of all was the Rittmeister Manfred Albert Freiherr von Richthofen, a 'Junker' cavalryman who first became an aerial observer and gunner, then, with the help of his mentor Boelcke, a pilot. Germany's leading World War I flying ace with eighty 'kills' to his name, the Red Baron finally lost his legendary concentration, flew too close to the ground for too long and was hit by a British .303 inch (7.7 mm) bullet. Shot through the heart, perhaps by a Canadian flyer, perhaps by an Australian machine-gunner, his focus returned in those last seconds. After landing his ungainly, bright-red Fokker Dreidekker, he spoke a few words to the first soldier who reached the scene, then died. Such is the stuff of heroes. They buried von Richthofen with full military pomp and circumstance. Australian soldiers carrying reversed rifles acted as the honour guard.

The *Peanuts* cartoons of Charles Schulz made at least the *idea* of the Red Baron familiar to later generations. Perched on top of his kennel in the clouds, the beagle air ace, Snoopy, strains forward, goggled, with leather helmet and scarf flying behind in the wind from the propeller wash of his imagined Sopwith Camel biplane. Suddenly, he spots a shadow moving across the ground: a rattle of guns, and the kennel is riddled with bullet holes. It's the Red Baron. 'Someday, someday, I'll get you Red Baron!' Dogs speak and a cardboard box or a kennel can be an antique fighter in the mind of any small child, cartoonist or thinking beagle.

Schulz wasn't a political cartoonist like Gary Trudeau, so *Peanuts* doesn't explore what happened after the real Red Baron fell from the skies. The book-reading Bavarian Hermann Goering enhanced his reputation by commanding the conspicuously painted planes of the Red Baron's 'Flying Circus' through the final phase of the war. He went on to great fame and notoriety as Hitler's Luftwaffe commander. Reichsmarshal Goering was also known as a virulent anti-Semite, a *bon vivant* and a major art thief. If he was a bibliophile in his young days, that certainly did not continue. The Nazis burned books and Goering is credited with the words: 'When I hear the word *culture*, I reach for my Browning.' Sentiments like that are still the stock in trade of extremist political thugs. He boasted to General Halder that he had set the Reichstag fire, the event the Nazis used to end any pretence of democratic rule. No longer the trim flyer shown in a photograph reproduced in *German War Birds*, Goering took potassium cyanide shortly before he was to be hanged at Spandau prison after being condemned in the Nuremberg war crimes trials.

Von Richthofen, like the World War II British general Bernard Law Montgomery, understood that the basic business of modern war is to kill the enemy. A trained hunter and careful tactician, he would identify the straggler or the less competent flyer in a formation, then pursue relentlessly till the targeted aircraft was in flames or diving out of control. Values have changed, and the fact that he ordered a small silver cup engraved with the number, the type of aeroplane and the date for each 'kill', is recorded without distaste in *The Red Knight of Germany*. The Red Baron followed Immelman's precept to stay high in the sky and keep the sun behind him. Allied flyers were warned to 'Beware the Hun in the sun'. There is a painting with that title in the Imperial War Museum. Before they took off, the Germans wished each other 'Hals und Beinbruch' (loosely: break an arm and a leg), believing that this would bring them good luck.

At the beginning of the conflict, primitive aeroplanes like the German Rumpler-Taube, the French Farman and the British BE2 were used solely for reconnaissance. The graceful, bird-like Taube, which can be seen at the Boeing Air and Space Museum in Seattle, first flew in 1910 and (lacking the aileron flaps that planes still use today) was steered by wires that warped the flexible wings, the archaic turning mechanism used by the 1903 Wright Flyer. The BE2 could achieve a maximum speed of 116 kilometres per hour in level flight. By 1917, the allied Sopwith Camel could reach speeds in excess of 190 kilometres an hour, depending on which particular engine was fitted. This was a classic arms race. Flying at 222 kilometres an hour, the British SE5A of 1917 was much quicker and easier to fly than the brightly coloured Tripes of the Flying Circus, which were soon replaced by the faster Fokker Eindekker, a high-wing monoplane. These speeds must have seemed enormous to former infantry and cavalry officers flying in open cockpits.

The Red Baron's triplane and the Camel biplane that may have ended his life were hard to fly and even more terrifying to land. Though they did destroy more than 1200 enemy aircraft, at least 300 Camel pilots misjudged and died in accidents. The main reason that the Tripe and the Camel were both dangerous and deadly was that they were powered by almost identical Rotary engines, based on an original French design, the Le Rhône. Though they were not used after World War I, the Rotaries were light and delivered a lot of power, making the aircraft very manoeuvrable. Many examples survive in various collections and, as mentioned earlier, the engine of the Red Baron's last fighter can be seen at the Imperial War Museum.

Front-on, the propeller of a Rotary aircraft engine is bolted directly to the hub (crankcase) of what looks a bit like a wagon wheel with nine large, finned spokes (the cylinders) and no rim. This configuration seems very odd to any contemporary observer with a knowledge of motors. Much like in an automobile, the type of internal combustion power plant we find in a modern, light aeroplane has pistons that slide up and down (or back and forth) in the cylinders of a fixed alloy block to turn the rapidly rotating crankshaft and attached propeller. This was also true for the Hispano Suiza-powered SE5A of 1917, and for Hermann Goering's Mercedes-engined Halberstadt. Conversely, the central, immovable crankshaft of a Le Rhône or Oberusel Rotary was attached to the Sopwith Camel or Fokker Dreidekker airframe so that the heavy engine, pistons, cylinders and propeller spun around it (and thus the forward axis of the plane itself) as a single unit.

The gyroscopic effect resulting from this 'flywheel-like' arrangement had major consequences for the handling of these lightweight aeroplanes, causing both the Camel and the Tripe to turn much more easily to the right than to the left. This was used to advantage by experienced air warriors, but learning to fly one of these spirited and unpredictable beasts was no doubt helped by having the 'hands' and quick reflexes of a cavalry officer accustomed to excitable horses. Also, as the Rotary engine could run at only one speed (flat out at 1200 rpm for the early versions), the pilot had a 'blip' button that allowed him to cut and restart the motor while landing. Just think how comfortable you would feel as a passenger if stopping and starting the engines was the accepted procedure for bringing a jet airliner to earth.

The lightness of the Rotary power plant was achieved by dispensing with the need for both a separate flywheel and a sump full of heavy-duty oil that prevents the piston rods and crankshaft of a conventionally configured internal combustion engine from seizing up. That's the burnt-smelling gunk that drains out when your car goes in for an oil change. The trade-off in this power-to-weight ratio game was that the Rotaries were lubricated by a spray of castor oil mixed with the gasoline fuel. The downside of that strategy was that combustion was inefficient and caused the engine to spew out a lot of excess oil and gas, especially when the pilot was 'blipping' to reduce speed by turning the ignition on and off.

Castor oil was evidently in short supply at the end of World War I, causing significant difficulties for the German air force. Young children of my generation dreaded castor oil as the universal, nauseating, household treatment for constipation. That would not have been a problem for World War I pilots as, apart from the stress and excitement of battle, they would have been breathing in a lot of partially burned castor oil—which is, incidentally, also carcinogenic. Most of them didn't live long enough to develop cancer. The life expectancy of a new fighter pilot in 1917 could be measured in weeks.

Charles Schulz must have seen Howard Hughes's 1930 movie *Hell's Angels*, which killed three of the World War I veteran flyers and stunt pilots that he hired to recreate these spectacular aerial melees. Though shot in black and white, the movie makes it obvious why the planes were painted in bright, distinctive colours. The pilots often had only an instant to decide whether they had an enemy in their sights and take the shot. *Hell's Angels* is still worth viewing today for anyone interested in the history of both film and aerial combat. Hughes flew the final dogfight scene himself, crashed and was slightly injured. This aerial accident was the first of several that may have contributed to his later, famously eccentric behaviour. Was it the term dogfight that caused Charles Schulz to make the Snoopy connection? Beagles aren't all that pugnacious, but the reflective Snoopy image just isn't suited to a snappy terrier or a heavy-duty Rottweiler.

A long-running military debate about the manoeuvrability of multi-wing aircraft versus the speed of single-wing fighters had largely been resolved by 1939, though some smaller countries like Norway and Belgium were still flying British

Gloster Gladiator biplanes, and (in 1941) the antiquated
Fairey Swordfish 'string-bag' biplanes of the Royal Navy
sunk the battleship *Bismark*. Two squadrons of Royal Air
Force (RAF) Gladiators that flew in France at the outset of
hostilities were easy prey for the faster monoplanes of the
Luftwaffe. Fortunately for the future of western civilisation,
the RAF had settled its front-line defence strategy on the
Spitfires and Hurricanes that were to defeat Goering's air
armadas in the 1940 Battle of Britain. What mattered in that
war was speed, weight of armaments and rate of climb. The
streamlined, predominantly metal aircraft were too fast to
indulge in the spiralling dogfights of the 1917–18 wood, wire
and fabric era. The capacity to fire through the propeller no
longer mattered as multiple heavy machine guns or cannons
were mounted in the sturdy wings.

At the outset of this first aerial war, the pilots of Taubes
and BE2s fired pistols at each other. The next generation of
two-seaters carried an observer with a rifle, then a swivelling
machine gun that could shoot safely to the back or side of the
aircraft. By 1917, courtesy of Anton Fokker's interruptor
gear, The Red Baron's Dreidekker was fitted with two
forward-firing 'Spandau' (Maxim) machine guns. Called
Spandau because they were made in the factory town where
Goering was later to meet his destiny, they were essentially
identical to the allied Vicker's gun that, whether or not it was
the weapon used by the Canadian Captain Arthur 'Roy'
Brown in his Sopwith Camel or by the Australian Sergeant
Cedric Popkin on the ground, ended von Richthofen's life.

When Manfred, Roy or Cedric depressed the trigger
button, the firing pin was released to strike the detonator on

one of the 250 to 550 bullets carried in the belt feed. Even when fired in short bursts, a machine gun soon becomes very hot, from both the heat of combustion and the metal-on-metal friction of the speeding bullet. The cumbersome, trench-warfare version of the Vicker's gun used by Cedric Popkin surrounded the barrel with a bulky, fluted water jacket connected via a rubber hose to a separate can that condensed the steam from the boiling coolant. The weapon lacked the fixed platform of a tank or an aircraft, so six men were needed to maintain it, as well as its ammunition, condenser and tripod mount, in the muddy, shell-cratered terrain.

Because of both the weight and the problem associated with a water feed in a twisting and turning dogfighter, the Maxim water jacket was ventilated with holes or louvres in the aerial versions of the Spandau and Vickers machine guns. Mounted directly in front of the pilot, they were cooled by the air wash from the spinning propeller. The rapidly turning, finned cylinders of the rotaries that powered the Sopwith Camel and the Dreidekker cooled themselves. On the other hand, the business end of the 1917 Hispano-Suiza V8-powered SE5A fighter is dominated by coolant-filled radiators located directly in front of the engine. Like the 'Hissos', as they were known, the World War II V12 Rolls-Royce Merlins in the Mustangs and Spitfires were equipped with automobile-like radiators carrying a mixture of water and 30 per cent glycol, the anti-freeze that we use in cars today.

While the sealed, bubble cockpit of a late-model Spitfire or Mustang could become very hot when flying above the clouds in direct sunlight, staying cool is not likely to have been a problem for the World War I aces aloft in their sparsely equipped, open cockpits. The mandatory neck scarves were best made of silk so that they would not chafe with the frequent head turning necessary to check for the presence of the enemy. They also helped stop the flying castor oil from the engine dribbling down the pilots' backs. *The Red Knight of Germany* describes Von Richthofen's uniform as 'soaked with grease, oil, sweat and mud'. The 'Snoopy' goggles were essential to keep the oil out of their eyes and to protect against the blowback from the hammering guns firing right in front of their faces. British flyers smeared their faces with the whale grease issued to prevent trench foot, with the result that, when the pilots removed their goggles after landing, they could look like racoons or chimney sweeps. In winter, the flyers dressed warmly and often rather eclectically, wearing a variety of heavy coats, fur-lined boots, gloves and so forth over uniforms or more informal but warm underclothing.

All pilots fear fire. The engine, machine guns, pilot and unarmed fuel tank of the Sopwith Camel were close together in the front third of the aircraft. Yet it was only late in the war that the Americans introduced the familiar leather flying suit. The heavy clothing provided some protection, but there was still little hope of survival in a burning wood-and-fabric aeroplane. Though World War II flyers carried parachutes, many of those who lived after a cockpit fire were horribly disfigured.

Flying propeller-driven fighter planes required fast reaction times and physical strength. This was a young person's game and, with the exception of a few Russian women in World War II, combat flying was an activity restricted to young men. So far as I'm aware, only two pilots fought right through both world wars: Biggles and his chum Algy. The invention of a World War I Pilot Officer writing as Captain WE Johns, the upper-crust James Bigglesworth and his friend Algernon Lacy were familiar to just about every small boy of my generation who lived in the old British Commonwealth. These high-spirited adventures were favourite Christmas and birthday presents. Published between 1932 and 1970, more than ninety books took the doughty duo from Sopwith Camels to Spitfires and beyond. In order to add an element of youth, later stories also introduced Ginger Hebblethwaite and Bertie Lissie as key, additional characters

Captain Johns was no literary giant. The one rule he had learned was not to repeat verbs. The paragraphs with dialogue read something like: 'said Biggles', 'quoth Biggles', 'Biggles expostulated', 'groaned Biggles'. However, he could tell a good 'boys' own' story that enthralled the rather naïve kids of my pre-television childhood. The female of the species got an occasional mention in the Biggles series, but the very sketchy romantic involvement was incidental to the action. However, WE Johns wasn't a misogynist. When in 1941 the British Government asked him to write similar tales for girls, he introduced the eighteen-year-old Pilot Officer Joan Worralson and her friend Betty Lovell. Their flying adventures in a total of eleven 'Worrals' books were equally popular, ran on for a decade, and evidently solved overnight a wartime recruitment problem for the Women's Auxiliary Air Force (the WAAFs).

The gallant Captain Johns was a man of his time; his books can be massively politically incorrect by today's standards. When the more open and permissive 1960s and 1970s came along, Biggles, Algy, Ginger and Bertie became standard comedy targets for a while, but this soon passed as those who had read these 'ripping yarns' aged and dropped out of the broader audience. First-hand Biggles knowledge is now a province of nostalgia buffs, collectors and geriatrics.

Biggles' legendary and continuing enemy, the monocled von Stalheim, clearly had some of the Red Baron's attributes, though he was much more durable and lived well into the second half of the twentieth century. Fiction offers an alternative universe, and it's certainly not beyond the bounds of literary licence to speculate about possible outcomes if the Red Baron had also survived the 'Great' War. What would have become of Manfred von Richthofen if he had not taken a bullet through the heart and had, instead, endured as long as the imagined von Stalheim? One thing that's for sure is that Manfred would have buried his beloved Great Dane, Moritz, who often slept on his bed. Might he have married, continued his Junker line, and now have Olympic equestrian or fashion-designer great grandchildren listed in the Almanac de Gotha? Perhaps not: unlike his younger brother and fellow Flying Circus ace Lothar, an energetic womaniser, he is not known to have had much interest in the opposite sex.

Other than for the politically motivated removal of the brown-shirt leader Ernst Roehm, not being a ladies' man was never a major problem in the Nazi party. A living Red Baron would have been too much of a public figure, too valuable a propaganda tool to be allowed to disappear without trace in the Germany of 1933 to 1945. Deutschland's top, surviving air ace was Ernst Udet, with sixty-two 'kills'. Though no fanatic, he was a professional flyer and, when recruited to the

Nazi Luftwaffe by Hermann Goering, was instrumental in the development of the screaming Stuka dive bomber. He became a Luftwaffe General but, with the failure of the Battle of Britain, Goering managed to convince Hitler that Udet was to blame. Knowing his likely fate, Udet shot himself in 1940 while talking on the telephone to his then mistress. Manfred von Richthofen was a great tactician. Could he have won the Battle of Britain for Germany and changed history?

Even if he had remained in Hitler's good books through the early stages of World War II, von Richthofen might, as a prominent, aristocratic member of the Prussian military caste, have been at great risk after the failed assassination plot led by Claus von Stauffenberg. Had he avoided the choice of taking poison given to Erwin Rommel or being shot like von Stauffenberg, he could well have survived the Nazi Götterdämmerung and the equivalent of the gasoline fire that consumed the remains of Hitler, his propaganda minister Joseph Goebbels and their wives on the blasted terrain outside the Berlin bunker where they spent their final days. If Manfred had lived and returned to his home in Silesia after 1945, he would have died under another totalitarian, but now communist, regime.

Emigration is the one path that may have allowed the Red Baron to avoid any involvement with the Nazis. The dire economic conditions in post-World War I Germany could have led him to boost the family fortunes by accepting an invitation from Howard Hughes to play himself in *Hell's Angels'* dogfight sequence. The Red Baron was a versatile, though not outstanding, pilot who, if he had managed to keep flying, would have had no problem with the relatively docile Curtis biplane that displays the name 'Rittmeister von Richthofen' very prominently in Hughes's epic movie. Maybe Hughes would have given the real Red Baron a more prominent role, though he probably lacked the thespian skills of Lucien Privel who played Baron von Kranz, the German lead in *Hell's Angels*. The von Richthofen character has a speaking part in *The Blue Max*, a much later film about World War I German war birds, but as this was made in 1966 he would have been too old to function in other than an advisory role.

Enjoying a wealthy lifestyle and meeting Hollywood luminaries like Sam Goldwyn and Luis B Mayer, Manfred might have remembered his best friend of Flying Circus days, the Jewish ace Werner Voss, and decided that the poverty and emerging anti-Semitism in Germany were not for him. Like the ancestor of Charles Schulz, he could have chosen to spend the rest of his life in the United States, perhaps to become established in the film industry. With his classical, Teutonic cast of features and a suitably sinister accent, von Richthofen would have been a possibility to play von Stalheim in a hypothetical 1930s Biggles movie (or serial) for Saturday afternoon kids' matinees. On the other hand, these were still early days in Hollywood, and anyone who had successfully commanded a fighter squadron that was constantly moving its base and flying under very varied and hostile conditions has at least some of the characteristics needed to be a movie director.

Might a newly sensitive, outward-looking Freddy von R have added an Oscar to the collection of little silver cups stored in a cupboard back in Europe, perhaps we would now be talking about the historic films and eccentricities of both von Stroheim and von Richthofen. The (probably) ersatz aristocrat Erich von Stroheim played the relatively sympathetic German commander in Jean Renoir's *Grand Illusion*, an anti-war movie that drew on Renoir's experience as a World War I cavalry officer and reconnaissance pilot, and directed a number of films including *Greed* and *Foolish Wives*. Jean Renoir, who moved to Hollywood when he fled the Nazis, was the son and biographer of the painter and sculptor Pierre-Auguste Renoir. The 1937 *Grand Illusion* was banned in both Hitler's Germany and Mussolini's Italy. If he had established his acting credentials in *Hell's Angels*, Von Richthofen could also have been an obvious choice to play alongside von Stroheim in the 1932 flying movie *The Lost Squadron*.

The only von Richthofen pilgrimage that we ever made was to the grave of his distant relative, Frieda. The young Emma Maria Frieda Freiin von Richthofen first married a British academic, then dropped him for D (David) H Lawrence of *Lady Chatterley's Lover* fame. They both lived for a short time in Thirroul, south of Sydney. Despite the brevity of his stay, Lawrence's *Kangaroo* is the best evocation I've read of the longing for a simplistic, authoritarian leader that periodically blights the Australian political landscape. David died in France and Freida brought his ashes back to the only house they had ever owned, a simple cabin in the scented, peaceful forest outside Taos, New Mexico. Frieda lived on for a further quarter century and, in 1956, was buried in front of the modest memorial to Lawrence that stands in the garden. Determined that David's remains would stay in the grounds of their home and not be transported to some distant pantheon by reverential acolytes, she dumped his ashes into the wet concrete that solidified to become the shrine's altar. There he remains.

Both Manfred and David were close to their mothers. Perhaps the hypothetical, more culturally aware 'movie baron' might have decided to make a film of *Sons and Lovers*. Linked by exile and social background, the North American von Richthofens may have become good friends and Freddy could have eventually retired to New Mexico to be near Frieda. The hills near Taos are a fitting place for a hunter at peace, whether the quarry was a fragile flying machine or the vagaries of the human spirit. In their different ways, the latter describes both David and Frieda.

The real Manfred von Richthofen was twenty-five when he fell. Is it a tribute to those who died to give them imagined lives? So many of the best and bravest were gone forever. What might they have achieved? If it is acceptable, in *The Plot Against America*, for Philip Roth to have the aviator hero Charles Lindbergh defeat Franklin D Rockefeller in 1940 to become an anti-Semitic, pro-Nazi, US president, then I suppose we can also allow some little exploration of the human potential that was lost on the killing fields of the western front. How different our world could have been if that ill-judged and terrible war had never occurred. What have we missed in not experiencing the later song cycles of George Butterworth (killed at Pozières in 1916), the Welsh bard Hedd Wyn's (killed at Ypres in 1917) Nobel Prize-winning poems from the 1930s, or the great plays that Wilfred Owen (dead at twenty-five, just before the armistice) penned in his fourth and fifth decades?

The true literary legacy of World War I is extraordinary, numbing grief. Among the most moving books that I've read are *Sunset Song*, the first volume of Lewis Grassic Gibbons' trilogy *A Scots Quair*; and Vera Brittain's *Testament of Youth*. Gibbons was the nom de plume of novelist James Leslie Mitchell, who served as a clerk in the World War I Royal Flying Corps. Based on a friend and real farm wife, his heroine and dominant character, Chris Guthrie (Mrs Tavendale) of Blawearie, experienced the loss of her husband and many of the men and boys of her acquaintance who served in the North Highlander regiment.

Vera Brittain, the mother of former Labour cabinet minister Shirley Williams (Baroness Williams of Crosby), describes how, as a young, middle-class, North-of-England woman, between 1914 and 1918 she experienced the death or severe injury of every young male she was related to or knew closely. The compelling, contemporary novel *Birdsong* by Sebastian Faulks deals with the enormous psychiatric toll, while the 'shell shock' aspect of World War I is also at the heart of Grassic Gibbons' *Sunset Song*. Then there's Pat Barker's magnificent novel *Regeneration*, which, taking a true story as its base, explores the poet Siegfried Sassoon's emotional and intellectual rejection of this insane and useless conflict. Sassoon was a decorated and courageous officer. If, like Ewan Tavendale in *A Scots Quair*, he had been a private soldier, he would have been shot for desertion by the British authorities. Anyone who hasn't read, in particular, *Testament of Youth* and *Regeneration* should take the time to do so.

The spirit in 1914 Europe was that this would be a jolly good fight, a chance for chivalry and valour. Part-time soldiering was a feature of life in the first decades of the twentieth century and large numbers of men were rapidly mobilised for military service. They marched off in high spirits and were cheered on by their women and male relatives who were too young or too old to fight. Nobody expected that the conflict would bog down so cruelly in the mud and blood of France and Flanders. Even the generals had forgotten the lesson of battles like Gettysburg: that it is suicidal to send massed infantry against heavily armed, entrenched positions. There were no machine guns at Gettysburg, just repeating rifles and revolvers. Perhaps remembering the slaughter of the Civil War, the Americans came late to the 1914–18 conflict. When the world was called on to fight again between 1939 and 1945, there were few illusions about what was likely to happen between the opposing forces. It wasn't World War I but the later Spanish Civil War that prepared us for the horror of burning, bombed, civilian neighbourhoods, then whole cities in flames. War would never again be a circumscribed event where 'ignorant armies clash by night' (as described by Matthew Arnold in his 1867 poem *Dover Beach*).

After thousands of years, this first 'Great' European war also marked the effective end of the horse soldier, both as a significant fighting unit and as a cultural icon of personal, particularly aristocratic, courage. The last successful cavalry charge of World War I was not by sabre-wielding heavy dragoons led by the younger sons of lords or barons but the 1918 opportunistic dash by plebeian mounted infantry of the Australian Light Horse at Beersheeba in the open terrain of the Middle East. Armed with rifles and bayonets, they used the speed of their horses to surprise and bypass the Turkish defenders. The muscle and spirit of a well-bred, carefully trained charger was replaced by the much more controlled horsepower of one or other type of oil-guzzling engine. After 1918, members of cavalry regiments rode in armoured cars, tanks and helicopters where, unlike the experience of the horse-borne warrior, they ran the risk of burning to death. Magnificently outfitted horse guards are now used only for ceremonial occasions. Mounted police patrol parks and control unruly crowds, but they don't carry swords or flag-tipped lances.

Manfred von Richthofen entered the conflict with all the élan and high spirits of a young Uhlan officer. He quickly became bored with his marginalised role as a mounted scout, hated the mud of Verdun and decided to become a flyer. Hermann Goering came across from the Cycle Corps, though his desire to leave the damp earth was evidently triggered by an attack of rheumatism. Of those who died in the air, the lucky ones like the Red Baron took a bullet. Others fell agonisingly in their flaming machines before mercifully hitting the ground. The fiery prelude that led ultimately to both the ovens of Auschwitz/Birkenau and the incineration of Dresden was played out in the fields and skies of the western front.

The Red Baron was a much decorated and celebrated aviator, though that could also be said of Hermann Goering. Nobody doubts the bravery of those who fought, but anything positive that came out of World War I was greatly outweighed by the immense suffering and the effective destruction of a whole civilised, but in some senses naïve, way of life. Even when World War I ground to its final, bloody end, the economic punishment inflicted on Germany by the terms of the Treaty of Versailles led directly to the humiliation and degradation that contributed to the rise of Nazism. By the latter half of the 1930s, the former infantry corporal Adolf Hitler and his followers were again preparing to 'let slip the dogs of war', both against the citizens of other countries and the Jewish, gypsy and intellectually or physically compromised members of their own population.

This time we did understand what needed to be done after the cessation of hostilities, and the Marshall Plan of 1947 ensured that there would be no further conflagration in Western Europe. Though there was a standoff with the USSR through the Cold War, the leaders of the two superpowers were fully aware of the possible consequences and chose not to fight. Hopefully we still carry that lesson with us today, although the Iraq disaster has made many of us wonder whether our current heads of government have any sense of history. Lest we forget.

Beacons

On Alexander the Great and the
Hatteras Light

On a fine, clear morning, we looked
out from an open terrace set high in the facade of Egypt's
magnificent, modernist Biblioteca Alexandrina. The encir-
cling, light grey outer wall of the Biblioteca stretched beneath
us. Carved with symbols, hieroglyphs and letters both archaic
and modern, it symbolises the complexity, beauty and continuity of language and the written word. Beyond that lies the
contrast of a busy, dirty highway, packed with commuters,
trucks and the commercial life of a modern city. What most
draws the eye and the mind that it serves, though, is the
peaceful harbour and breakwater of the ancient port of
Alexandria, then the dark blue of the Mediterranean beyond.

According to the director, Ismael Seregeldin, stone
building blocks from the ruined Great Library of Alexandria
were directly in front of us, though now under water.
Founded by Ptolemy I, the Great Library was a beacon of

light and learning in the ancient world. Among other functions, it served as the collection of record for that extraordinary creativity that we associate with Athenian democracy. Presaging our era when one copy of every book published is held in one or other National Library, Ptolemy III ordered that all manuscripts and scrolls brought to Alexandria must be surrendered while they were copied by scribes. The papers of Archimedes and Euclid were stored there, more than a hundred scholars were in full-time residence and it is thought that the library functioned a little like a university of today.

The location of the world's first great library in North Africa reflects the fact that early Alexandria was as much Greek as Egyptian in character. Philip II of Macedon unified the Greek city states, then his son, the warrior Alexander the Great, disseminated Hellenic power and culture throughout the Mediterranean region and beyond. Among Alexander's many achievements were the liberation of Egypt from Persian rule in 332 BCE and, after Homer appeared to him in a dream when he was camped opposite the tiny island of Pharos, the founding of the city that bears Alexander's name. Dead at thirty-three, he didn't even see the first buildings on the site. According to legend, his body was preserved in a barrel of honey so that it could be taken to Alexandria, where it remained on display in a glass sarcophagus for many years.

Alexandria passed through Roman (30 BCE), Islamic (642 CE), French (1798), then British (1882) control, before becoming modern Egypt's second city and main port. During July 1942, Erwin Rommel's Afrika Corps pushed the British Eighth Army led by Claude Auchinleck back across the desert

to El Alamein, a small, coastal railway junction 102 kilometres west of Alexandria. In October, now commanded by
Bernard Montgomery, the Eighth Army's tank and infantry
brigades, including Australian, New Zealand and Greek
battalions, fought the second battle of El Alamein that ended
the Nazi presence in North Africa. Neither Alexandria nor
Cairo ever came under German control.

The atmosphere of turmoil and intrigue in wartime
Alexandria burned itself into the mind of then resident
Laurence Durrell, to emerge later in his illuminating
masterpiece *The Alexandria Quartet*. The novels *Justine*, *Clea*,
Mountolive and *Balthazar* tell the same story of love, politics,
deception and betrayal from the different, individual
perspectives, a technique that was completely new to most
of his readers, including me. In retrospect, Durrell shines
some light on how centuries of humiliation and colonisation
have fed the current anger and divisiveness in this part of
the world. We may, however, get more insight on that
particular set of issues by reflecting on the ways that globalisation, mass tourism and satellite television are likely to
have affected the claustrophobic, traditional society that
is so beautifully and affectionately described in Naguib
Mahfouz's *Cairo Trilogy*.

The tensions in Alexandria were palpable during our brief
2006 visit. A fatal, well-planned terrorist bomb attack in the
Egyptian Red Sea resort town of Dahab put the security
forces on high alert. Baton-armed, helmeted riot police and
their support vehicles were very obvious in the streets. As we
all have come to realise, the struggle between Egypt's efforts
to sustain a modern, tolerant, secular state and the extremes

of Islamic fundamentalism are not likely to be resolved easily, or in the near future. These tides in human perceptions are flowing in very different directions. They meet in Alexandria where, in some neighbourhoods, we saw men and women, in light dresses with their hair flowing free, eating and drinking coffee together in roadside cafes while, in other places, the scene was men only and the few women we saw were covered and anonymous to the outside world.

Opened in 2002 with support from many Arab nations and the dedicated patronage of Egypt's First Lady, Her Excellency Suzanne Mubarak, the circular Biblioteca Alexandrina represents a concerted effort to make this proud and ancient city once more a major centre of international culture. It is already serving as an important meeting place for representatives from societies that differ greatly in wealth, belief systems and opportunities but are dedicated to promoting global amity and prosperity. The Biblioteca is just one example of the continuing intellectual, scientific and economic resurgence of a moderate, forward-looking Islam that those who live in the west hear far too little about. Along with the spectacular economic developments in the Emirates, the bright lights include the construction of major technology and university facilities in Qatar and the aggressive pursuit of a spectrum of business, educational and industry initiatives in Malaysia.

With the Mediterranean and the old port as a background, director Ismael Seregeldin spoke passionately about the aims and the achievements to date of the modern library and reminded us of the contributions that ancient Alexandria and the subsequent Islamic intellectual culture made to human progress over its first thousand years. Touring the facility, we saw school children participating in 'hands on' science programs and sensed the same excitement and curiosity in these kids and their young teachers that we would find in western interactive science museums. We marvelled at the technology, including immersion in a 'virtual reality' experience, a seemingly solid representation of a massive human skull and brain that floated as patterns of light in front of us.

All the visitors on that open terrace had recently travelled far and, in the balance between breakfast coffee and the effects of jet lag, were expecting a long day inside talking and listening to conference presentations. Some were contemplating their own performance in a cross-disciplinary format aimed at developing insights for promoting biotechnology research, application and development in the less wealthy regions of the globe. Serious stuff, but now, on that sunlit space overlooking the cobalt-blue Mediterranean, we had the small indulgence of a few minutes to step back and dream a little of those who went before, long before: to imagine what might have been if the early record of experiment, thought and wisdom accumulated in the ancient library had been valued and preserved intact through more than two millennia to our own times. Sadly, that respect for knowledge and human intellectual activity is too rarely sustained through human history. My mind wandering, I experienced a brief sense of concern for the Biblioteca itself as I thought of narrow, fundamentalist perspectives that are too often blinkered by their exclusive focus on the dazzling luminescence of the one, 'true' light. How is it that the words of charismatic visionaries that were intended as bright truths in very different times can be used so perversely by those whose authority is threatened as an excuse for repression, violence and mass, indiscriminate murder?

The Ptolemaic library was first burned, though accidentally, by Julius Caesar in 48 BCE, and may have been finally suppressed under the Roman emperor Theodosius 1 in 391 CE. One version has it that the Christian Patriarch Theophilus used the Emperor's decree as a justification for displaying, then burning, the 'pagan' mysteries of the library. Another has it that the manuscripts came to be so little valued that they were fed into the fires that heated the water in bath-houses. Curious, isn't it? Pagan seems to be identified with beauty and reason in some people's minds, while infinitely painstaking and honest work can be so carelessly destroyed. We see the same thing over and over through human history, with the persecution of intellectuals and book burning in Nazi Germany, the damage to Chinese historical sites during Mao's Cultural Revolution and the destruction of the Bamiyan Buddhas by the Taliban being among the more recent iterations.

Though we have a fragmentary understanding of what was held in ancient Alexandria, most of this extraordinarily valuable resource was probably lost forever. The destruction of the library marks the beginning of the Dark Ages, the thousand-year interval when, at least in Europe, the authoritarian imposition of dogma and a dull scholasticism focused the adventure of the human spirit more on creating great works of art and architectural masterpieces like the Gothic cathedrals than on the pursuit of reason and open enquiry. For a millennium, the light of curiosity that drives science and discovery was dimmed. More than twelve centuries passed before René Descartes could confidently write '*Cogito ergo sum*': I think therefore I am.

Much of what was preserved from the ancient Greek and Egyptian cultures contributed to the early intellectual flowering in the Arab world which, in turn, seemed to falter as the European Reformation and the Renaissance freed enquiring minds from the tight control of ecclesiastical suppression. Such a loss will not occur again. The collections of the new Biblioteca are electronically archived and stored on four continents. The custodians are from cultures with widely different belief systems and all, presumably, share an attachment to truth and evidence-based reality.

Out on the raised breakwater of Alexandria's old port, beyond the site of the ancient library, a square, mediaeval Turkish fort is thought to stand on the base of what was one of the seven wonders of the ancient world, the marble Pharos, or great lighthouse of Alexandria. The small island of that name seen by Alexander the Great has now been incorporated into the harbour wall. The lighthouse, completed around 285 BCE, was somewhere between eighty-five and 140 metres high. Built on a square base with thirty metre sides, it had three levels topped by an eight-sided tower and a cylinder that carried the cupola where the flame burned. The source of light was probably a smoking fire fed by rushes lifted to the top in a sort of dumb waiter, a labour-intensive system that would also have required constant supervision to ensure that high winds did not extinguish the blaze. Some accounts relate that a polished, metal mirror concentrated the light from the flame at night, or the sun during the day, allowing the beacon to be seen from a much greater distance. As with modern lighthouses, the Pharos was a tourist attraction and visitors could climb to the top. Eventually neglected due to the diminished importance of the port of Alexandria, then destroyed by earthquakes, it stood for more than 1500 years and was greatly admired.

The purpose of a lighthouse is to guide mariners to a safe harbour or to warn them of shoals, rocks or a dangerous coastline. Being at least 600 years too late, we were not able to scale the Pharos of Alexandria but, like many reasonably fit vacationers, we've never been able to resist the challenge of scrambling to the top of any lighthouse that's open to public view. My most memorable experience of this type is climbing the 286 stairs of North Carolina's Cape Hatteras Light. Painted in a striking black and white spiral, this iconic nineteenth-century lighthouse reaches some sixty metres from base to spire on the sandy strip of the Outer Banks. Beautiful and romantic though they are, the Outer Banks are known for an extremely dangerous coastline that is justifiably called 'The Graveyard of the Atlantic'.

The Outer Banks mark the point where the northbound Gulf Stream and cold currents from the Arctic run head on into one another, creating the treacherous, shifting sands of the Diamond Shoals that stretch more than twenty-two kilometres from Cape Hatteras into the Atlantic. The area has claimed hundreds of ships. Many foundered when, driven by wind and water, they struck on the sandy shallows. Others were sunk by German U-boats, principally during the first half of 1942. Heading north under tow for repairs the revolutionary but barely seaworthy Civil War ironclad

USS *Monitor* was lost in a severe 1862 storm off Cape Hatteras. The wreck of the *Monitor* was located in 1973 and is a protected historic site.

These narrow, barrier islands look eastward across the turbulent Atlantic towards Europe and Africa and west to the placid inter-coastal waterway that stretches the length of much of the eastern United States. Though, unlike Alexandria, the Outer Banks were never a haven for philosophers and thinkers (unless on vacation), they were the site of a bold experiment that changed the human reality forever. In 1903, two bicycle manufacturers from Ohio—Orville and Wilbur Wright—first achieved a few metres of powered flight at Kitty Hawk, about halfway up the main island of Hatteras. A century later we take it for granted that we can travel from Nags Head on the Outer Banks to London, England, then to Alexandria, Egypt, in twenty-four to forty-eight hours.

The present lighthouse at Cape Hatteras is the third in a series. The first Hatteras Light used twelve whale-oil lamps with convex reflectors but, especially under stormy conditions, these were barely visible to shipping. This situation changed dramatically in 1822 when the French physicist Augustine Fresnel invented the lens that bears his name. Anyone who has ever climbed a lighthouse will have marvelled at the massive (some stand three to four metres high), elaborately shaped glass they find at the top. Numerous concentric rings act as prisms to focus the light that a centre drum section bends to shine though the bull's-eye shaped lens. The net result is that a 100-watt light bulb can give a concentrated beam with a brightness equivalent to the power of 6000 candles, and is visible for more than thirty-two kilometres.

The heavy lens is mounted on a wheeled base, which rotates on a circular rail to give the repeated flashes that (depending on the interval) constitute the lighthouse 'characteristic' read by passing mariners. The Fresnel lens of the current Hatteras Light was originally turned by a slowly descending weight that the keepers would need to wind back to the top of the tower at no doubt too regular intervals. The light source itself changed over the years from whale oil to kerosene to electricity. Now all lighthouse functions are automated, and those lighthouse keeper's cottages that are not too remote are often holiday homes or small museums. Ships rely on satellite navigation, though it must still be comforting on a storm-tossed night to see those flashing, penetrating lights that symbolise a connection to terra firma. Fresnel lenses continue to find new applications, including concentration of the sun's rays to provide a source of renewable energy.

Both the Hatteras Light and the twenty-metre Ocracoke Light (built in 1820) on the charming small island to the south were damaged during the Civil War when Confederate troops, fleeing after the battle of Fort Hatteras, took the lenses with them. The second Hatteras Light was not worth salvaging, but the third (and current) tower was completed in 1870, with the new Fresnel lens first being lit on 16 December of that year. Then wind, water and weather took their toll, till the progressive, natural degradation of this sandy coastline brought the lighthouse base perilously close to the breaking waves. For a time it was replaced by a steel tower, but a change in the pattern of erosion and the desire to preserve what is a striking and beautiful monument led to it being returned to use in January 1950.

The next, and perhaps most extraordinary, step in the saga was to shift this sixty-plus-metre brick structure more than 870 metres back from the Atlantic Ocean, a much debated move that was completed in 1999. We can only be grateful that, in this case, 'the ayes had it'. The distant, then close, view of sea and sand and the black-and-white barbershop pole of the Hatteras Light lifts the human spirit. Nobody who sees it can feel anything other than delight—unless they've been partying too hard on a Nags Head or Duck (a town, not a bird) vacation and their companions insist on climbing to the top.

Sand islands like the Outer Banks enjoy a place in history and in the affections of those who visit or live there, but they can have no permanent status in geological time. As it is, human intervention has greatly influenced the present form of this tenuous environment. Since the post-Depression era of Franklin Roosevelt's Work Projects Administration (the WPA), an eighty-kilometre dune built up with timbers and sand has slowed the natural process of erosion. The highway that runs the length of Hatteras also provides a further source of stability, limiting the natural, westward movement of the sandy landscape.

Even the most modest predictions of ocean rise due to global warming have serious implications for the Outer Banks. Ocracoke is, on average, less than two metres above sea level. Flooding of the low lying salt marshes and erosion around inlets will rapidly change the topography and, with similar processes happening at many sites along the nation's coastlines, there would be no resources available to halt such changes. As it is, the United States Congress already forbids the construction of sea walls to halt processes of natural beach erosion. Recent experience suggests that severe storms and hurricanes are likely to cause increasing damage to man-made structures all along the eastern US seaboard.

Water is far from the sole invader that threatens human communities. Through recorded history, hilltop beacon fires have warned of attacks by pirates, raiding parties seeking slaves and loot, and the arrival of hostile fleets and armies. Local communities in England and Wales supplied fuel and maintained the pyres so that they could be lit at very short notice. Names like Ilminster Beacon, Dunkery Beacon and the Brecon Beacons remind us of this early alarm system. In mediaeval times, the beacon hills were organised with three separate stone hearths, or iron tripods. One fire warned of imminent danger, two that invasion was likely, and three that the enemy had already landed.

The Romans lit signal fires along the length of northern England's Hadrian's Wall. The Great Wall of China, which stretched for more than 6000 kilometres with the intent of excluding Mongol cavalry, was never contiguous in desert regions but consisted instead of forts that communicated by means of predetermined patterns of smoke and fire. Every kid of my pre-television generation who spent an occasional Saturday afternoon in the canvas seats at the local movie house knew about Indian smoke signals from cowboy movies featuring Tom Mix, Hopalong Cassidy, the masked Lone Ranger and his partner Tonto. The distant puffs of smoke were always accompanied by an ominous drumming sound: with the exception of the somewhat inarticulate Tonto, we had no doubt who the bad, uncivilised guys were.

As kids in sub-tropical Brisbane, we sang: 'Fire in the mountains, run boys run, you with the red coat follow with the drum'. Or was it 'gun'? Where did that come from? Red Coats were British soldiers of the eighteenth century and the earlier part of the nineteenth century. 'Boys' could refer to children or to groups like the 'White Boys', a ground-roots Irish Catholic movement that resisted control by absentee landlords and Protestant England by burning fields and houses. Alternatively, do the words celebrate a hilltop beacon fire lit to alert the Red Coats to follow the 'Oak Boys' or the 'Protestant Boys', descendants of colonising lowland Scots brought across at the time of Oliver Cromwell? The fire that this seemingly light-hearted little rhyme celebrates may reflect a dark, exploitative and at times murderous reality. It dips into a deep past that links bright, leaping flames set in high places with danger.

One group's revolutionary patriot is another's evil terrorist. Whatever we call them, the most dangerous among us burn with a bright, fierce, uncompromising light that illuminates a single, clear purpose. From the destruction of the Great Library, through to the mediaeval pyres lit by narrow zealots like Torquemada and his opposite numbers in the Protestant countries, to the horror of Pol Pot in Cambodia to 11 September 2001 and the whole mess of diffuse and state-sponsored terrorism of today, it is only when we work out how to clear the flammable brushwood, to damp down the infernos and turn this enormous human energy and clarity of purpose in positive directions that we achieve harmony, a true state of grace.

Beacons for human progress like the United Nations, global treaties that preserve the air and the oceans, accords that respond to the rights and needs of persecuted women and children, and the international courts of justice that bring criminals at every level to account do not always burn as brightly as they should. We need to work out how to enhance such positive flames, to clean the lenses that focus the twin beams of the intellect and moral consciousness that protect the human family.

Tall Ships, Black Gangs, 'Bully' Wars

HMS Temeraire *to USS* Cole

For three centuries, from Henry VIII's *Mary Rose* of 1510 to the Napoleonic wars of 1803–15, the peak of naval power was represented by tall-masted wooden ships that moved through the water by the power of wind on canvas and delivered their *force de frappe* as devastating, multi-cannon broadsides. The recently re-discovered wreck of the *Mary Rose* was lifted from the ocean floor and taken to be conserved at Portsmouth's Royal Navy Dockyard, long the home of Admiral Nelson's flagship at Trafalgar (1805), the 'first rate' (more than 100 guns) HMS *Victory* that is available for all of us to visit and climb over. The other great reminder of Trafalgar is JMW Turner's evocative 1838 painting *The Fighting Temeraire*. The massive three-decker that followed the *Victory* in line of battle is shown under tow by a steam tug, en route to her final berth at a breaker's yard in Rotherhithe on Thames.

At the height of the conflict the *Temeraire*'s captain, the hot-tempered Eliab Harvey, pulled his 98-gun 'second rate' out of the *Victory*'s wake to put her alongside the French *Redoubtable*, which she quickly reduced to a splintered wreck. The *Redoubtable* subsequently went to the bottom but both British vessels, despite substantial damage, were later returned to duty. Admiral Sir Eliab Harvey, a renowned eccentric and heavy gambler, served alternately as a member of parliament and a naval officer and lived on for another twenty-five years, long enough to see auxiliary paddle steamers towing naval sailing ships out to sea from Portsmouth Harbour.

The 'fighting' *Temeraire*'s bold move at Trafalgar was too late, though, to save the life of Horatio, Lord Nelson, who fell to a musket ball fired by a sharpshooter high in the *Redoubtable*'s rigging. Refrigeration was yet to be invented, so Nelson's body was brought back to England in a cask of brandy (or was it rum?) to be displayed at Greenwich, then interred in London's St Paul's Cathedral. There's a tale that some of the crew surreptitiously drilled a hole in the barrel and drank the preserving fluid. 'Tapping the admiral' is an illicit drinking alternative to 'splicing the main brace', which refers to the double issue of rum given to British sailors as they worked to reinforce the rigging before battle. Traditions change, albeit more slowly than ships, and the last official rum ration was served in 1970 on HMS *Endymion*.

A 2005 contest run by BBC's Radio 4 that attracted 119 000 voters chose *The Fighting Temeraire* as the greatest painting in Britain. It normally hangs in London's National Gallery and is readily viewed on-line. To the extraordinary aesthetic appeal that characterises Turner's misty seascapes is added the pictorial tension between the primitive, squat, black paddle-tug belching fire and smoke from her high funnels, together with the towering, archaic grace of the soon to be dismantled *Temeraire* and the brilliant sunset, and the total effect is both unforgettable and deeply moving. Turner's imagery, which relied heavily on artistic licence, subtly captures the essence of the industrial revolution: the sudden transition from a long, stable, romanticised past to a rapidly changing, more pragmatic and dirtier (at least from the aspect of air quality) future. By 1860, the innovative iron-clad HMS *Warrior*, which rests at Portsmouth with the *Mary Rose* and the *Victory*, was powered by both sail and steam.

Oak-hulled sailing ships like the *Victory* and the *Temeraire* made some use of the coal and metals that were to define this next stage of sea travel and marine warfare. Apart from the copper sheets that protected the *Victory*'s bottom from growth by marine organisms and thus enhanced her speed, most of the metal she carried was in the iron ballast that kept her upright; in the heavy guns and the myriad of smaller armaments that protected her; in the shot, anchors, ranges and cooking utensils that served her needs; and in the nails, bolts and pins that held her together and helped the crew work the complex rigging. The *Victory* also carried fifty-one tonnes of coal for essential activities like fuelling the galley fires, blacksmithing, heating shot in battle and melting pitch for caulking the joints in her deck and hull.

By the end of the nineteenth century, the navies of the world were comprised of coal-fired, iron-hulled steamships. Admiral Dewey's flagship at the battle of Manila Bay (1898), the twin-screw, second-class battleship (or armoured cruiser)

USS *Olympia*, is preserved at the Philadelphia's Independence Seaport Museum. Arguably as historically important and unique a maritime relic as the *Victory*, the *Olympia* (built in 1891) is basically a construct of brass and riveted iron rather than the aluminium, titanium and welded steel that are characteristic of a modern warship. She has bunkers to carry more than 816 tonnes of coal and her two, tall funnels reflect the fact that, as she moved through the water, the exhaust from her furnaces rose as a constant plume of black smoke and soot. Viewed from a distance at her berth in the Delaware River, though, she doesn't look too different from those mid-to-late twentieth-century destroyers and cruisers that still carried big guns rather than guided missiles as their main armament.

While her four slender smoke stacks give away that this is a ship from an earlier era, that same sense of being more or less in modern times applies as we sit back and watch one of the many movies made about the iconic first and last 1912 voyage of the White Star liner RMS *Titanic*. The upper-class passengers may have dressed and acted more formally, but the popping champagne corks, grand ballrooms, cabins and dining experience would be familiar to anyone who has recently taken an up-market ocean cruise. If, though, we could time-travel back to visit the 'below the stairs' engine rooms on the *Titanic*, what we would see there would be quite unlike the throbbing, massive, bright-painted diesel/electric power plants that propel the floating vacation palaces of today. The difference relates to the source of the energy that pushed the *Titanic* through the water at forty-one kilometres per hour, or twenty-two knots, a speed considerably in excess of that associated with a relaxed cruising experience. In the age of jumbo jets, massive passenger vessels no longer compete to secure the Blue Riband awarded for making the fastest transatlantic voyage. What drove the *Titanic* can be summed up by four factors: steam, coal, human ambition and muscle. Much of that maritime reality and lifestyle is now—along with the basics of many people-intensive, old technologies—almost forgotten.

The stokers who made up the 'black gangs' in coal-fired steam ships were tough, highly skilled labourers. Apart from having their clothes, skin and lungs constantly impregnated with black dust as they loaded, shovelled, spread and trimmed huge quantities of coal during their four-hour shifts, they monitored the airflow in the furnaces, dealt with the clinkers (lumps of impurities) and, under the direction of the engineering officers, responded to the engine telegraphs from the bridge. Even when the ship was at rest, some were on duty to feed the fires that served the ancillary engines powering the electric dynamos, the compressors for the ice-making machinery and refrigeration plant, and the condensers that turned salt water into fresh.

Given the signal 'full speed ahead', the watch engineer of the *Titanic* would adjust the settings on the Kilroy Stoking Indicator to ensure the supply of high-pressure steam to drive the ship's three engines and propellers. Only some of the twenty-nine boilers heated by 159 furnaces were required and needed feeding at any one time, information that was passed on to the black gang by the loud bell of the Kilroy. Full speed on the *Titanic* meant moving something like 408 tonnes of coal in a 24-hour period. Ringing with monotonous regularity day and night, this electric drill sergeant controlled the working lives of the stokers as inexorably as the drumbeat that set the rhythm for ancient galley slaves. At least the rowers had a portal to the wind and sea and were not at risk of anthracosis. The immediate precursors of the black gang, the 'top-men' more than fifty-five metres up in the rigging of the *Victory*, would have had the ultimate clean air experience.

Being in the black gang on a north-Atlantic liner like the *Titanic* would have meant staying warm in a cold climate. Shovelling tonnes of coal and living with blazing furnaces must have been an even harsher reality in the heat of the tropics. While wealthy travellers and colonial administrators could purchase 'POSH' (port-out, starboard home) first-class tickets so that they could experience a shaded, cooler passage on the British/India (BI) and Peninsular and Oriental (P&O) boats heading through the Suez Canal to India, the Far East and Australia, that option was hardly open to the stokers.

Toiling in the bowels of the ship, the members of the black gang were also the least likely to survive if, as in the case of the *Titanic*, the hull's integrity was suddenly destroyed. The *Titanic*'s officer of the watch tried to steer around the legendary iceberg but, striking at a speed of more than twenty knots, the vessel was breached below the waterline in six separate places. Though recent examination has established that the physical damage was much less extensive than previously thought, the location of the various penetrations fatally compromised the effectiveness of the *Titanic*'s system of watertight bulkheads. Apart from collision, the other major event that can quickly send a large, iron ship to the bottom occurs in wartime, with the destructive force of a blast from a floating mine, or a torpedo fired from a stalking submarine.

Resting at anchor in Havana Harbour on a fine, quiet evening in February 1898, the more than 5000 kilograms of gunpowder in the magazines of the USS *Maine*, the younger, half-sister ship of the *Olympia*, exploded and blew the vessel out of the water before she settled and sank in the shallow waters. The forward third, which housed most of the crew, was effectively destroyed and 266 died. The officer's quarters were in the rear of the vessel and the majority, including the captain and six of the seven engineers, survived. The black gang fared less well. Of the fifty-plus listed as 'coal passers' or 'firemen', only four lived to tell the tale.

Following a board of enquiry finding that a mine had been set under the *Maine*'s powder magazines, public anger directed at the Spanish who controlled Cuba was a major triggering event for the brief Spanish–American war of April to August 1898. The linked call to arms of 'Cuba libre' and 'Remember the *Maine*' roused the nation. Culminating with the battle of Manila Bay and the capture of that other Spanish territory, the Philippines, this marked the United States' first major adventure as an intercontinental power. The USA also gained control of Puerto Rico and Guam.

It was a simpler and, in some ways, less brutal time than we know today. Serious, educated people could still talk of a 'splendid little war' at the end of the nineteenth century, that distant mirror before humanity had experienced the charnel house of the Western Front. Are we any less deluded now? Televised 'entertainment war' allows us to watch 'smart bombs' blow up 'enemy' targets and, perhaps, incinerate hundreds of unseen and often innocent people. Sitting with our TV dinners, we feel a warm glow of national pride at the state of our advanced technology and the professionalism of our warriors. The slaughter of the Civil War battles like Shiloh (1862) and Gettysburg (1863) was remembered with horror by the veterans who were still alive in 1898 but, as we saw again in Iraq, those who have not endured battle can often commit a nation to war with a naïve sense of arrogant optimism and adventure.

Unlike contemporary politicians, enthusiastic patriots like Colonel Theodore Roosevelt and his slouch-hatted Rough Riders put their own lives at risk and gloried in the fight as they charged up San Juan Hill, an action that greatly enhanced Teddy's public career. Assistant secretary of the navy in the McKinley administration at the time of the *Maine* disaster, Teddy Roosevelt, went on to be the twenty-sixth president. He was certainly the most colourful of the twentieth-century Republican incumbents, although, unlike some of his successors, he was also a progressive reformer who broke the monopoly control of the industrial 'robber barons'. An emphatic man, he was known for exclaiming 'great Scott' and 'bully' (superb, wonderful), an expression that endures in Roosevelt's description of the presidential 'bully pulpit'. He also gave us the Teddy Bear, and Monty Python fans will remember the whimsical cartoon of his very recognisable, walrus-moustachioed head opening its mouth for a bee to emerge.

But what really happened to the *Maine*? The initial, hastily convened board of enquiry had little technical expertise.

A much more thorough 1911 investigation built an elaborate cofferdam around the wreck so that the water could be pumped out and the damage examined at close quarters. Photographs taken at that time confirmed that the explosion had effectively blown the ship in two. This second committee, meeting long after the heat had gone out of the Spanish–US confrontation, supported the earlier conclusion that the trigger causing the mass of gunpowder in the *Maine*'s forward magazines to explode was a mine placed externally. However, no residual mine or torpedo casings were ever found, and the argument was based on the fact that a few (and one in particular) of the ship's bottom plates were bent inwards.

Some of the technical experts present in 1911 disagreed with the majority conclusion and, in 1976, 'The Father of the Nuclear Navy', Admiral Hyman B Rickover, argued in his book *How the Battleship* Maine *Was Destroyed* that the real cause was a fire started by spontaneous combustion in a poorly ventilated coal bunker. Bunker A-16 was separated by a 6.3 millimetre steel plate from a reserve magazine that contained tanks of brown powder and black saluting powder—an arrangement that, even then, was considered dangerous by some naval personnel. Furthermore, as the ship was at rest, the pipe that ventilated A-16 to one of the *Maine*'s smoke stacks would have lacked the 'exhaust-convection' effect supplied by the hot gases from the boiler furnaces. Once the minor, ancillary magazine went up, the rest followed and the ship was destroyed from within.

The US battleships of that era, including the *Olympia* and the *Maine*, had the coal bunkers aligned adjacent to, and outboard of, the magazines, a measure intended to give some protection against attack by the newly developed torpedo boats. Unlike oil, coal stored in the absence of any external source of combustion can be a very dangerous product. When freshly broken, it oxidises in a process that generates heat, while coal that contains a lot of iron pyrites can heat and burst into flame when wet. Newly mined coal may also give off methane, which can seep out of the bunkers and accumulate in the below-deck areas of a ship. That problem was solved by placing those familiar comma-shaped deck ventilators fore and aft, thus allowing a constant airflow to wash out any potentially incendiary gas.

Bunker fires were well known in the steam-powered vessels of the late nineteenth and early twentieth centuries. Though no such occurrences had been recorded for the *Maine*, the *Olympia* had three events of this type between 1895 and 1898. Extinguishing a stubborn bunker fire was one of the many challenges faced by the black gang in the epic voyage of the USS *Oregon* as she raced from San Francisco, around the tip of South America and through the Straits of Magellan to play a part in the 1898 naval battle at Santiago de Cuba. She fired a shot across the bows of the Spanish cruiser *Cristobal Colon* (Christopher Columbus), thus convincing the captain to avoid the slaughter of his crew by ordering that the sea-cocks be opened and the ship scuttled. The *Colon* had run out of high-quality steaming coal and, slowed by burning an inferior grade, she was rapidly run to ground by the heavier *Oregon* and her sister ship the *Brooklyn*, a vessel that also had a history of coal-bunker fires.

From the time the *Titanic* left Southampton on 10 April, she had a persistent, smouldering fire in her number 10 coal bunker. Two stokers from each four-hour watch were detailed to try to deal with it. That the captain of a luxury transatlantic ocean liner like the *Titanic* would embark passengers and leave port with this problem seems incredible to us now, but the apparently cavalier attitude illustrates how common such occurrences were in coal-fired ships. One theory has it that the need to enlist the help of the New York City fire boats to extinguish a bunker fire that was marginally under control contributed to the high-speed dash that ended in the *Titanic*'s loss in the early hours of 15 April 1912. Other evidence suggests that the ship's steel plating over bunker number 10 may have been unduly brittle, an effect that could be related to overheating. Bunker 10 is located next to boiler rooms 5 and 6, the site of the greatest hull damage.

Less than a third of the 240 or so crew members listed as 'trimmer' or 'firemen', and not one of the twenty-six engineering officers, escaped from the *Titanic*. Some died quickly as the water flooded through the breached hull. Others continued to serve the remaining furnaces, allowing the pumps to help keep the ship afloat until their flames were also quenched by the steadily rising water levels. At least one of the electric generators functioned till the end. The electric lights were still burning as she went down.

The heavy loss of black-gang members compares with the survival of about two thirds of the sixty or more shown on the manifest as deck officers and crew. Both their physical location on the ship and the fact that the seamen would have been required to man the lifeboats presumably gave them a better chance than the engineers. In retrospect, a job that involved 'rearranging the deck chairs on the *Titanic*' may have had some positive aspects. Those lost included the captain, the first officer, the assistant surgeon and all of the eight band members who, as we all know, went down playing the hymn 'Nearer My God to Thee'. Fire and ice: it is thought that the bunker coal continued to burn after the *Titanic* upended, broke in half and sank, the only source of warmth in those freezing waters.

Unlike the newly in-service *Titanic*, the *Maine* operated under strict naval discipline and long had the reputation of being a tightly run ship. Evidence presented at the first board of enquiry detailed both the careful routine that was followed to minimise the possibility of a bunker fire and the lack of any indication that there was such a problem. Even so, the external-mine versus internal-bunker-fire debate has not been, and probably never will be, resolved, although it does seem unlikely that the Spanish authorities in Cuba would have wanted to provide any excuse for the United States to attack them. It also seems a bit far-fetched to think that the rebels—who, like the Americans, were trying to get the Spanish out of Cuba—would have been sufficiently devious and well-organised to set the mine themselves so that the blame would fall on the Spaniards. If, on the other hand, burning coal provided the fuse for the *Maine* explosion, it raises the intriguing thought that the Spanish–American conflict was the first of what might very loosely be described as the United States' 'fossil hydrocarbon wars'.

Moves to deny essential hydrocarbons to a potential enemy set up the inevitable train of events that led the

United States into World War II. Concerned about Japan's brutal incursion into China, Teddy's distant cousin Franklin Delano Roosevelt, the thirty-second president, acted in concert with Britain and the Netherlands on 1 August 1941 to impose a total oil embargo. At that stage the USA were supplying more than 80 per cent of Japan's oil, with much of the rest coming from the Dutch East Indies, now part of Indonesia. In the certain knowledge that their powerful navy could not function long without oil, the military establishment of imperial Japan regarded the joint embargo as an unstated declaration of war. Seeking to establish their hegemony in the Pacific and to secure continued access to the oil pumped in the colonial territories to their south, the Japanese launched their coordinated attacks on British-controlled Hong Kong and the US fleet at Pearl Harbor on 7 December 1941, the date that FDR described as 'a day that will live in infamy'. The invasion to secure the oil fields of Burma began on 11 December, and by early 1942 the Japanese were established in northern New Guinea en route to Australia and had taken the Dutch East Indies.

The combination of 'embedded' journalists and video has ensured that America's two very recent but more limited 'hydrocarbon wars' are familiar to all of us. Over the past twenty years, the First Gulf War launched under the leadership of President George H Bush (Bush 41) to drive Saddam Hussein out of Kuwait, then the Iraq invasion instigated by President George W Bush (Bush 43), has shown how any real or perceived threat to the continued availability of the most valued of all hydrocarbons, oil, has potential for spontaneous political combustion. In the future, tensions will inevitably be exacerbated as the supply becomes more limited due to the progressive, rapid consumption of this non-renewable resource. Though other more esoteric and high-minded political motives may be stated, future conflicts are likely to be driven by the need to compete for key, increasingly scarce natural resources.

If the more extreme of the global warming scenarios turn out to be true, the need for arable (or even dry) land and water will become a very prominent cause of war. There is nothing new in this. The desire for territory and empire has long provided a major justification for military expansion and colonialist adventurism, stretching back through the Romans to Alexander the Great, the Persians and earlier times. Operation Barbarossa, Hitler's invasion of the Soviet Union, was intended to gain territory (*Lebensraum*) and to ensure the continued supply of oil from the Caucasus to fuel his war machine. The never-ending, and seemingly unresolvable, conflict between the Palestinians and Israel is as much about land and water as anything else.

From the 1773 Boston tea party that helped bring on the War of Independence to the various confrontations involving ships and sailors before the British–American war of 1812–14, to the loss of the *Maine*, events involving ships have often provided a trigger for subsequent conflicts. Most people have seen the iconic pictures and the movie re-creations of the bombed USS *Arizona*'s burning bridge and listing mast as she settled to the shallow bottom of Pearl Harbor. A quarter of a century earlier, the loss of 128 American lives when the fast transatlantic Cunard liner the RMS *Lusitania* was torpedoed in May 1915 by a German submarine helped develop the climate of opinion that allowed the twenty-eighth president, Woodrow Wilson, to bring the United States into World War I.

The sinking of the *Lusitania* near the end of her 202nd (eastward) voyage was associated with a relatively small primary explosion (the torpedo), followed by a much larger secondary blast. The cause of the latter is debated. While it seems that the *Lusitania* was illegally carrying at least some military supplies, it is not known how much, as the cargo manifest was probably falsified. Thousands of unexploded, standard British Army issue rifle bullets have been found in a forward hold of the wreck—although, even if these had gone off, they would not have caused all that much damage. A second possibility is that the torpedo ignited a mix of air and residual dust from the 6000 tonnes of coal that had been in the almost empty bunkers. Whether powdered hydrocarbons or clandestine munitions provided the explosive force, the rapid, severe listing that resulted from the starboard bow being blown off greatly enhanced the loss of life. Marine investigators are still pursuing the question, so the nature of the destructive explosion that destroyed the *Lusitania* may someday be resolved.

What was claimed at the time to be two separate August 1964 attacks by North Vietnamese gunboats on the destroyers USS *Maddox* and USS *C. Turner Joy* in the Gulf of Tonkin was used by the thirty-sixth president, Lyndon B Johnson, to persuade the Congress to pass the Gulf of Tonkin Resolutions that provided the legal justification for involving US troops in the disastrous Vietnam adventure. Just how much the President knew at the time is not clear, but it is now obvious that the first incident was relatively minor and involved only the *Maddox*, while the second was, at best, imagined and had the American gunners firing at phantom targets.

There was, however, no doubt about the reality of the October 2000 attack on the guided missile destroyer USS *Cole* in Aden Harbour, Yemen, when two suicide bombers rammed her with a small boat carrying 400 to 700 pounds of explosive. The ship survived the twelve-metre hole blown in her side and was returned to the United States, but seventeen of her crew were killed and a further thirty-nine were wounded. This occurred on President Clinton's watch and, though the event primed political and popular thinking for the later interventions in Afghanistan and Iraq by the soon-to-be-elected Bush 43 administration, no direct military action was taken against the core terrorist al-Qa'ida organisation at that time. In retrospect, given the events of 11 September 2001, that lack of a significant response looks to have been a mistake.

Warships have long projected the power of great nations and, as a consequence, have also served as remote targets for hostile forces. President Theodore Roosevelt thought it 'bully' as he farewelled the 1907 'Great White Fleet' of warships that steamed 69 230 kilometres and visited six continents to show the US stars and stripes. From August to September 1908, they visited Sydney, Melbourne and Albany. Consisting of four squadrons, there were sixteen 'battle wagons' (including the second USS *Maine*) that had been commissioned since the end of the Spanish–American war as well as the USS *Olympia*. Apart from her role as flagship of the US Asiatic Squadron at the Battle of Manila Bay, the *Olympia* served in the Atlantic and the Russian Arctic in World War I and brought back the body of America's unknown soldier in 1922. She was decommissioned soon after that but was preserved as a relic and in 1957 was returned to her 1898 configuration. If you visit her in Philadelphia, look closely at the engine room and also at the mark on the deck showing where Admiral Dewey gave his famous order: 'You may fire when you are ready, Gridley'. That polite instruction sounds a little quaint, or even vaguely Pythonesque, today, but war is war and the consequences of Captain Charles Gridley's subsequent commands were no joke for the recipients of the exploding projectiles from the *Olympia*'s four eight inch and ten five inch guns.

Up till the time of the Great White Fleet, and the naval expansion program launched by Kaiser Wilhelm II in Germany, Britain's Royal Navy was the undoubted mistress of the seas. As kids in the former colonies that federated in 1901 to become Australia, we learned to sing: 'When Britain first, at Heaven's command, arose from out the azure main … Rule Britannia! Britannia rule the waves: Britons never, never, never will be slaves.' It's a great, jingoistic song that's still sung at the last night of the BBC Prom Concerts (popular classics) in London. Australia in the first half of the twentieth century, with its large land mass and its tiny population, felt protected by British naval power, an illusion that was shattered forever when, on 10 December 1941, Japanese torpedo bombers sunk both the battleship HMS *Prince of Wales* and the heavy cruiser HMS *Repulse* off the coast of Malaya. In August of that year, the towering, armoured superstructure of the *Prince of Wales* had provided the imposing back-drop for a meeting between Winston Churchill and Franklin Roosevelt. The loss of these two state-of-the-art capital ships was a bitter blow in the darkest time of World War II and, along with Pearl Harbor, provided the lesson that the days when massive, fast, floating battle wagons dominated the oceans were effectively over.

The ancient Red Sea port of Aden where the USS *Cole* was attacked is now the second city of the Republic of Yemen. By the late eighteenth century Aden had much deteriorated and in 1839 it was taken over and colonised by Britain to become, at least for a time, a strategically important outpost of Empire. In 1877, Queen Victoria was proclaimed Empress of India, the jewel in the British imperial crown. After the Suez Canal opened in 1869, both the warships of the Royal Navy and the civilian liners that linked Britain to India and more distant British possessions in Burma, Malaya and Hong Kong stopped for fresh coal and boiler water at Aden. This was the era when imperial Britain really did 'rule the waves' and drew on all the resources of her scattered empire. Locally recruited Yemenis and Somalis served on British merchant and navy vessels, generally as stokers. Though their special immigrant status in the United Kingdom stipulated that they could work only in shipping-related activities, many of their descendants have been assimilated into modern Britain.

The Aden coaling station was on the island of Perim and, although it seems very odd when we think in terms of the distribution of fossil hydrocarbon-based wealth and power today, the high-quality steaming coal was carried out from Britain (often from Wales) on sailing ships. Working these colliers was a grim business, especially as they travelled the long route via the Cape of Good Hope. Now Aden has oil tanks. The switch to oil-fired steam turbines in ships everywhere began with the Queen Elizabeth-class super-dreadnoughts (mega-battleships) commissioned between 1913 and 1915 when Winston Churchill was First Lord of the Admiralty, equivalent to the US Secretary of the Navy. Apart from the fact that they were faster than their coal-fired German counterparts, the considerably higher density of oil greatly increased their steaming range. Furthermore, it was much easier to take on fuel by pumping rather than by hauling sacks of coal; the need for a large black gang was removed, there was less smoke to make the ships visible on the horizon, and there were no more smouldering bunker fires. The same change soon followed in the merchant marine, with, for instance, the *Lusitania*'s sister liner the Blue Riband-holder *Mauretania* converting in 1921 to an oil burner.

It is possible that the realisation that oil would be an enormously important resource for the future was a consideration in the secretive Anglo–French Sykes-Picot agreement of 1915–16 that led to both countries maintaining a controlling role in the Middle East. Though the major discoveries in Iraq and Saudi Arabia came later, a fifth of the 1916 oil needs of the Royal Navy were already being supplied by the Anglo–Persian Oil Company. The post-1918 territorial carve-up, and the borders of Iraq, Syria and Iran drawn at that time, are central to the continuing problems in that part of the world. This was one disastrous example where, as great and powerful countries sometimes do, Britannia waived the rules—in this case of rational diplomacy and fairness to Prince Faisal's Arab forces which, recruited to the allied side by TE Lawrence, had helped defeat the Ottoman Turks. Such a major betrayal created a legacy of bitterness and enduring distrust.

The current political and religious tensions, exacerbated by terrorism and a climate of military confrontation, indicate that we will need to do something very different if there are to be positive outcomes in the Middle East. That reality will become even more pressing for those countries in the region that fail to seize the opportunity for modernisation offered by oil revenues. Their situation will only worsen as the supply inevitably declines and western countries develop the technologies that will enable them to become oil-independent.

Now, when we associate spontaneous combustion with hydrocarbons, our thoughts turn to oil-related political and terrorist activities that pose much more difficult issues than the relatively simple engineering problem of smouldering coal in ships' bunkers. In many cases they reflect the realities of the oil–power equation that seems often to facilitate regressive, rigid authoritarianism and the inevitable repression and violence that is associated with such a world-view. Remembering the *Maine*, we look back to what now seems a more innocent and less complex time. The fact, though, that European culture essentially self-destructed in the inferno of World War I should give us pause.

Are even those of us in the so-called advanced world any less susceptible to embarking on exercises of massive, collective stupidity? Perhaps the current catastrophic but limited conflagration in Iraq may have some positive benefit by reminding us that combining duplicitous propaganda, deliberate ignorance, blind political arrogance and force of arms is not likely to provide any useful solution to long-term problems. Can we finally accept that the days of militaristic fantasies and 'bully' wars are over, that we need instead to develop mechanisms based in evidence and reason that sustain global prosperity, equity and stability by ensuring mutual respect?

Firefighters

From gunpowder to Jet A fuel

Little boys want to grow up to be firefighters, an ambition that few of their parents take seriously as it usually fades early. Fire trucks and fire helmets are favourite toys. Small boys love engines of all types and descriptions, but the bright red, noisy fire engine is right at the top of the list. Among our eldest son Jim's first complex use of words was 'light ba/ba' to describe the flashing lights and the two-tone klaxon used by the fire trucks in Edinburgh, where he was born and lived for his first two years. Almost three decades later, his 2-year-old son Finn was more excited by sitting in the driver's seat of a retired Pumper donated by the local fire department than by anything else he discovered at the Memphis Children's Museum.

Fires have always been part of a more dangerous reality for scientists like me, especially in earlier days when the use of highly flammable solvents was much more widespread and less strictly controlled. I expect that most research workers of my age either know, or know of, someone who was severely injured, or even killed in a laboratory fire, even though such events, fortunately, are rare. There's always a fire extinguisher on the wall, everyone is supposed to know how to use it and, as on a ship, there are regular, compulsory fire and evacuation drills. That becomes even more important when, as is the case at St Jude Children's Research Hospital in Memphis, a high-powered medical research institute is part of a hospital with patient beds for very sick kids.

The fire station closest to St Jude is just across the street. The fire marshals and firefighters visit regularly, both in a planned way to check out that the correct standards are being observed and, more frequently, as an emergency response when the alarms go off. (During my time at St Jude, there were a few genuine fires, including a big one that was fortunately contained by the sprinklers. The problem then was flooding.) Mostly, though, the fire trucks are called out because of some glitch in the alarm systems, a fairly common occurrence as St Jude has a continuing building program, and things always go wrong in that transition phase when the controls for complex engineering structures and air-conditioning and heating systems are bedding down. When the alarm sounds and the loud 'Dr Red, Dr Red' announcement that's used in US hospitals comes over the loudspeakers, everyone evacuates via the stairs and assembles at a set meeting point. We then wait till the firefighters have been able to check out the problem and are satisfied that it's safe to return. Suffering, or enjoying, the sudden and unexpected displacement of busy people, everyone mills around, chats and, as the Scots would say, generally 'skyves-off'.

Day-dreaming one day while watching the firefighters going about their business, I registered the fact that the ladder truck bore the prominent name 'Richard C Shadyac'. Dick Shadyac, who later headed the St Jude fund-raising organisation, is a good friend, a leading Washington lawyer and a stalwart member of the St Jude Board, the predominantly lay governing body that hires and fires the director, decides on the annual budget and so forth. St Jude makes an enormous contribution to the self-respect and financial well-being of Memphis, and either the municipal government or the fire department wanted to honour Dick for his services by giving him his own fire truck, so to speak. In Britain, he would have been elevated to the title 'Sir Richard', but a more likely award in the United States is to have a boulevard, a bridge, or some other structure bearing your name. I happened to mention Dick Shadyac's fire truck during the time that I was doing a lot of public speaking as the 1997 Australian of the Year. Someone in the Brisbane fire department picked up on that, and the result is that my name is on an aerial ladder platform truck that can reach some forty metres off the ground. It was pretty exciting going up in this very impressive 'cherry picker'.

That visit to Brisbane's Fortitude Valley fire station was the first time I'd been in a fire house for more than fifty years. When I was a small child, we occasionally went to see distant relatives in an old, multi-storey fire station in South Brisbane where the firemen's (there were no firewomen then) families lived in apartments that were 'over the store'. This was very unusual for the Brisbane of that era, where just about everyone grew up in detached, wooden houses set on large garden allotments. When there was an emergency, the father wrapped himself around a slippery pole that descended the height of the building, then slid three or four storeys to the ground where the fire trucks stood waiting: a short but dramatic commute. Looking tentatively through the hole in the floor, down the length of the pole, to the ground far below ended any ambition that I had to be a fireman.

One of the big 'kid' occasions every year was Guy Fawkes' night, the inherited celebration that commemorates the failure of the Catholic conspiracy to blow up King James I and England's parliament: 'Please to remember the 5th of November, gun powder, treason and plot'. Following a tradition that started by decree of the British government in 1606, the neighbourhood boys and girls dragged wood from everywhere to build a bonfire on vacant land not far from our house. The 'Guy' was made from sticks and old clothes, stuffed with straw and, like a Christian martyr of the sixteenth century, tied on top of the pyre. Pennies were saved to buy fire-crackers, 'Mount Vesuvius fountains', 'Catherine wheels' and sky rockets. The hoards of explosives could be enormous by the time the 5th arrived. Kids' bedrooms were bombs waiting to go off, just like the kegs of gunpowder that Guy Fawkes and his fellow plotters placed under Westminster Palace. Come to think of it, I don't recall that the local Catholic families participated. Reflecting the times, there was a certain lack of respect for the sensitivities of others.

As kids grew older, some of the cracker 'experiments' became downright dangerous. Simply blowing the lid off a tin can morphed into more sophisticated projectile systems. My dad's contemporary, 'Bunger' Clarke, was famous for the injury he suffered when a lit thunderer exploded while he was holding it in his mouth. I didn't know him before, but the lower part of his face seemed to protrude a bit and, from a kid's point of view, I got the sense his teeth might have been blown out and forward, which seems dentally unlikely. Living constantly with the reality of the great Australian dryness, the firefighters must have hated Guy Fawkes' night. Was it the firefighters, the lawyers, the insurance companies, the vigilantes of politically correctness or the 'save us from ourselves' types who won out in the end? I don't know, but the amateurishness and spontaneity are long gone; any such event is now strictly controlled in Australia. But it was great fun. The dads drank a beer or two, and the kids from five to fifty-five loved it.

The closest thing in the United States is Independence Day, although the fourth of July has a much more profound political significance than Guy Fawkes' night. When we were living in Memphis, some of the neighbours had skyrocket battles across our suburban street. This was illegal, but the local police patrol wasn't too interested and it was a minor social event after the dress-up parade in the morning. One of the streets was closed off, and a local fire truck led the procession of bikes and trikes, with the kids costumed to represent Old Glory, Uncle Sam and various other traditional symbols. The evening of the fourth was always marked by a big fireworks event on the banks of the

Mississippi, which many of the St Jude Hospital staff watched from the top of the parking garage.

Celebrating with fireworks goes back more than a thousand years, beginning with the Chinese, who are masters of this art and occasionally suffer major conflagrations when a fireworks factory goes up with a bang. Putting out fires became increasingly important as the human population increased and more people moved into towns with buildings that had a substantial component of wood, were often sealed with pitch and even continued the rural practice of having their roofs constructed out of thatched straw. In early seventeenth-century London, the parishes were given the responsibility of maintaining buckets, ladders and fire-hooks. These precautions proved to be totally ineffectual in 1666, when the Great Fire of London burned for five days and destroyed eighty-seven churches, including St Paul's Cathedral, and 13 200 houses.

The response to the Great Fire was to form insurance companies, which marked the fronts of their clients' houses with distinctive metal fire plaques. In the eighteenth century, the city of Philadelphia had a number of such companies. The problem was, though, if the wrong group got to your house first with their hand or horse-drawn pumper, on failing to see their plaque they just left the scene. Some of these plaques are still to be found in downtown Philadelphia's historic Elfreth's Alley. By the nineteenth century, we see the beginnings of the modern, urban fire department. The insurance companies realised that it puts an insured home at risk if the fire company lets the house next door that isn't occupied by one of their clients burn to the ground. The fire insurance companies of London came together in 1833 to form the London Fire Engine Establishment. There were soon eighty firefighters working out of thirteen stations.

Urban fire departments throughout the world are now staffed by paid, highly trained, full-time professionals. Both Australia and the United States also have, with their large land areas, a strong tradition of volunteer firefighters in rural areas. Apart from their importance in protecting lives and properties, these community organisations function as a 'glue' that helps to hold country towns together as they confront the ever-increasing economic and social threats to rural lifestyles and values. Kurt Vonnegut Jr, who was a volunteer fireman, celebrates an American version of this culture in his *God Bless You, Mr Rosewater*. The Australian volunteer fire service is probably a bit different, but then Vonnegut was a novelist, not a sociologist or an historian, so his cut on this way of life in the States may be as much imagination as fact.

The Victorian Country Fire Authority functions as a mix of paid professionals and some 60 000 volunteers operating 1500 fire tankers in 1200 brigades. They can also draw on the services of twenty to thirty water-bombing aircraft of various types. Like any form of firefighting, their activities put them in real danger. Two volunteer Victorian firefighters, a man and a woman, were killed during the severe summer bushfires of early 2006, in separate accidents involving fire trucks and uneven terrain. Further to the north, near Cootamundra New South Wales, the pilot of a water-bomber was lost when it developed a fault and crashed into a hill-side. Another volunteer died in the massive firestorm that erupted in East Gippsland in December 2006, and a group of

New Zealand professionals who had crossed the Tasman to help out were lucky to escape when their position was engulfed by fire. In 1998, five volunteers ranging in age from seventeen to forty-seven were trapped and died in the charred hulk of their truck at what had initially been considered a fairly minor conflagration near the town of Ballarat. A neighbouring fire company watched, unable to help, as they were hit by the inferno. Between 1993 and 2002, 185 US volunteer firefighters gave their lives.

Then there's the urban, professional firefighter as hero. Will any of us forget where we were on 11 September 2001? My plans for that morning involved flying from Memphis to Philadelphia to review research at the University of Pennsylvania's Abramson Cancer Center. It didn't happen. Driving down the freeway, the chilling news came through on National Public Radio's Morning Edition that another aircraft had crashed into the second tower of the World Trade Center. Though we had heard the breaking story of the first hit at breakfast, the details weren't clear and, like many, we assumed it was some sort of tragic accident involving a small plane. But lightning doesn't strike twice in the same place. The immediate reaction was that this had to be a coordinated suicide attack. Though the deep sadness and anger of that day return when I think back on it, the prevailing sense is more of tragedy and fear for the human condition. Is civilisation possible if people are prepared to promote the routine use of suicide and the slaughter of innocents to gain political ends? The perception that the world had changed, and changed for the worse, has been amply substantiated by subsequent events.

The story of the 343 New York City firefighters who died that day doing the job that they were trained for is burned into our collective consciousness. With no obvious thought for their own safety, they rushed into the burning, 110-storey buildings. All were men, though three women in other uniformed services were also lost. Among the victims was the Fire Department Chaplain, Father Mychal Judge OSM, sixty police officers and a diverse spectrum of human beings from investment bankers, lawyers and stock analysts at one end of the financial spectrum, to cleaners, waiters and security guards at the other.

A total of 16 000 emergency personnel and volunteers, including firefighters from the surrounding New York City stations and districts far and wide, then worked to the point of exhaustion for days in an atmosphere that was a toxic mixture of airborne, finely powdered concrete, glass, heavy metals (like mercury) and so forth that had come from the concrete shell, the steel framing, the floor coverings, the telephones, the executive chairs, the kitchens, the heating and air-conditioning systems, the art works, the miles of ducting and wiring, the elevators, the computers, the photocopiers and all the other elements of modern construction and business practice that formed the complex social and physical structure of the Twin Towers. 'World Trade Center Cough' was a common problem that required acute, and sometimes continuing, treatment. Others were psychologically damaged, and many have had to take early retirement on medical grounds.

Why did the towers collapse? Though they seemed massive and impregnable, the advanced (for their time) engineering techniques that had been used in their construction meant that they were 95 per cent air. Anyone who had been to the 'Windows on the World' restaurant on the 107th floor of the north tower will remember having to change elevators twice at intermediate levels, an innovation that greatly reduced the structural mass required for the mechanical task of lifting people 415 metres above the ground. Unlike in older skyscrapers, the architects and engineers used relatively little masonry. The strength of the buildings rested in the central core that housed the elevators and services, together with frequently repeated girders and columns of lightweight steel. That's why, when it all came crashing down, the pile of debris seemed surprisingly small, though it did amount to more than 1.3 million tonnes.

The major disaster resulted, though, from the ignition of about thirty-eight kilolitres of highly flammable Jet A fuel. The ready supply of oxygen through the breached outer walls, together with the combustible substance of the building, floor coverings, woodwork and the like, then added greatly to the inferno, with the temperature of the fireball reaching around 705 to 760 degrees Celsius. That was not sufficient to melt the steel framing, but it did decrease the tensile strength of the steel girders and columns by about half. In addition, the uneven distribution of both the loss of fireproofing and the fire itself led to the buckling and further deformation of the support structure. The weight of the concrete floors then caused the weakened perimeter columns to bend inwards, and the floors to lose their anchorage and start the progressive, cascading collapse that brought both buildings down with the incredible speed that we all saw on our TV screens. The Twin Towers stood for 56 and

102 minutes respectively after the initial impacts, and they fell in twelve seconds.

United Airlines flight 175 from Boston, which hit the south tower, included on their passenger list Sue Kim Hanson, Peter Burton Hanson and 2½-year-old Christine Lee Hanson. The little family was en route to Los Angeles to visit relatives. Peter was a musician, a committed gardener, a loving father, son and husband, and vice-president for sales at TimeTrade in Waltham, Massachusetts. Christine Lee was an enthusiastic gardener's helper. A talented scientist and a fine human being, Sue Kim was born in Los Angeles and had just completed her PhD degree at Boston University. When I had the privilege of giving the Sue Kim Hanson Lecture at Boston University four years later, her young colleagues and extended family members were still in deep grief at the sense of waste and terrible loss.

Over the months after 9/11, *The New York Times* ran pictures and short biographies of those who were killed, either in the buildings or in the planes. These 2819 human beings from 115 countries represented a cross-section of normal, decent American life, together with a subset of the international visitors who flock to the business and financial hub that is New York City. We read of scout leaders, community activists, committed parents, sons and daughters, sports fans, sportsmen, loving aunts, fun uncles, immigrants struggling to succeed, and all sorts of people who were just going about the business of their daily lives with no thought of harming anyone.

The circumstances of 9/11 forced us all to acknowledge that life is not safe, secure and certain. As all the great religions teach us, we are fated to be candles that may flame high for a brief interval but will soon burn down or be snuffed out. The 'ashes to ashes, dust to dust' of the Christian burial service is a very apt statement of the human condition. That realisation may always be close to the surface for those in dangerous jobs like firefighting. Yet while normal, happy, prosperous people in wealthy, western societies realise that there are dangers out there, most suppress that knowledge and keep it below the level of consciousness. The fact is, though, that human beings are rather fragile organisms who have heart attacks, suffer strokes, and become concussed when they fall off ladders or dive into murky water that is too shallow. More than a million people, including about 40 000 Americans, are killed every year in road accidents. Though a few may be suicides, the great majority set off with no thought that they will not survive the day. We expect to live when we board a jet airliner.

At least for a time, 9/11 also made us recognise that even the most prosperous depend for their safety on the hard work, dedication and professionalism of relatively poorly paid, hard-working men and women who fight fires, check baggage and passengers at airports, maintain jet engines and ailerons, drive ambulances, staff casualty departments, serve us drinks and hot meals on long flights, and police our cities and highways. Sometimes, because they commit acts of outstanding courage, these generally anonymous human beings receive a measure of brief, public recognition that generally takes the form of medals and affirmations. Often, any due honour is given by their own, as when police and the firefighters bury their colleagues who have died in the pursuit of their duty. There is a National Fallen Firefighter's Memorial in the grounds of the National Fire Academy at Emmitsburg, Maryland. None of this costs a great deal in the monetary sense and, though it is well meant, it's a financially cheap way for society to say thanks.

The families of the 9/11 victims were compensated, though the largest dollar amounts went to the survivors of those who were the highest earners. That is hardly likely to have included any of the firefighters. Our political leaders spoke with genuine emotion about the great courage of the firefighters, and those moving words honoured their sacrifice. But words are cheap, and they are often ephemeral. Were there substantial improvements in the working conditions and salaries of firefighters? Or did the fine and sombre pronouncements amount to no more than emotional grandstanding and overheated rhetoric? With the current form of economic globalisation that takes little account of the health, safety or equity of reward systems for those in many developing countries, working people in western democracies can't expect much more from their governments than deceptive rhetoric and occasional handouts at election time. The firefighters at least have the satisfaction of knowing that their jobs, dangerous though they may sometimes be, cannot be outsourced so that CEOs can earn even bigger incomes as we see improved returns on our investments.

The Hot Air Diet

On calories and fish fry

Gourmet dining, fine wines and plentiful, good food rate high among the true delights and privileges of our prosperous western lifestyle. An ideal vacation for some is to travel from vineyard to vineyard, restaurant to restaurant. Such indulgence is too often followed by just a little pain and guilt. Being thin is the image of our age, though most of us lack the genetic makeup to be cat burglars, marathon runners or supermodels and, in any case, we don't have all that much willpower. Even so, few who are steeped in contemporary values would take the 1930s University of Chicago President Robert M Hutchins' view that 'whenever I feel like exercising I lie down till the feeling passes'.

We are fundamentally conflicted when it comes to food. To some extent feasting and fasting are the sin and salvation of our protected, predominantly secular societies. Chefs, cooks and diet gurus are way up there on the celebrity charts. Watching Julia Child, Nigella or one or other of the 'Iron Chefs' slave over a hot stove (or wok) is terrific 'street theatre'. The successful diet types, on the other hand, play more the role of the confessor/priest than entertainer, but some of them present well and, like any good preacher, are generally heard with due attention and respect. Mireille Guiliano's *French Women Don't Get Fat* hit the top of every international bestseller list. That title is so evocative: great cuisine, fine wine, the best cheeses can somehow come together to give the style, the svelte slimness, the assured pelvic thrust that we all associate with Parisienne elegance. At least for people of my advanced years, the aristocratic Audrey Hepburn somehow comes into the equation. Though Ms Hepburn was from the Netherlands, she had what used to be called élan in spades—and was unbelievably thin.

There are overweight people in France, but a diet book called 'Fat, French and Fasting' would sell less well than Tom Vernon's *Fat Man on a Bicycle*. The former title would be unappealing and even a little insulting to those who have a weight problem, while the latter works because it conjures a sympathetic image that is, at the same time, slightly Chaplinesque. As we age, many of us end up with this problem of being 'a fat man inside a thin man who is doing his best to get out'. This complex and conflicted relationship with food is very much part of modern life. The late-nineteenth to mid-twentieth-century idea of a 'well set up man' pretty much went out with Mrs Beeton and recipes that started 'take two pounds of butter and a quart of cream'.

Every weekly newspaper has a restaurant guide. The menus may have little heart symbols to show the 'healthy' items, but the tiramisu and the 'death by chocolate' dessert can be hard to pass up if we've been following a raw carrot, prune juice and skinless chicken diet. Even your doctor will tell you that an occasional 'diet holiday' is a good idea. Health farms, calorie restriction programs and the like are staples for the glossy magazines. Publishers are delighted by the fact that, while we buy large numbers of colourful, attractively illustrated cookbooks, the latest, high-profile diet book is greeted with equal, if not greater, enthusiasm. Most diet books are cookbooks in disguise, as much of their considerable volume is used to provide a range of 'health' recipes.

Food is important and people have been writing about it forever. Cookbooks have been around for more than a thousand years and some even go back to Roman times. Diet books seem to be more recent: William Banting's 1864 title *Letter On Corpulence* would probably not appeal to a contemporary publisher. My first encounter with a diet book was when I dragged the 1952 publication *Diet Does It* by Gayelord Hauser from a dusty bookcase in my grandparents' house. It wasn't that I was headed for anorexia nervosa or any similar dreaded disease; it was a combination of inertia, adolescent boredom and nothing to read on a too hot, enervating Sunday afternoon that led me to browse Hauser's magnum opus. The blue, plain, cloth-bound volume belonged to my very thin, cheerful aunt, Frances Byford. It was a family joke that Francie talked the 'salad, tomato and lettuce' talk but was easily seduced into dietary turpitude when confronted with a large slice of cream cake or the like.

My dear aunt is long departed but, unpacking a box of books after a recent move, we realised that we still had her copy of *Diet Does It*. Looking at it again as a veteran of the diet book reader scene, I realised that, though the book's style is dated, Gayelord Hauser was a pioneer. Because of my need to counter a bad cholesterol profile, Penny and I have accumulated a range of diet books over the years. The problem only really came under control, though, when I started taking a daily dose of one of the statins, those magic drugs developed from the discoveries of Joe Goldstein and Mike Brown at the University of Texas. As with most long-term drug treatments, I realise that I'm part of a massive, human, Lipitor experiment. Anyway it's good for scientists to have the guinea pig experience, especially if you're (hopefully) part of the positive response group.

Some diet books are inevitably better than others. Though they may emphasise this or that strategy—pasta/no pasta, fat/no fat, eggs and bacon/egg whites only—the general message is that having a balanced dietary intake that includes some fruit and vegetables, getting a little regular exercise, drinking enough fluid, and not over indulging in food or alcohol is a reasonable way to live. This was pretty much what Gayelord Hauser wrote way back then. Though most diets are sensible (if boring), some are more than a bit mad and even contribute to the deaths of their enthusiastic proponents. Speculation, unsubstantiated claims and just plain hot air are not exactly unknown on the diet guru scene.

We have to eat and drink. Our brains and bodies are heat engines that must have continuing access to oxygen and energy supplied in the form of blood glucose. Depending on the size of an individual's fat, glycogen (carbohydrate) and protein (muscle) stores, we can mobilise these reserves and survive for several weeks without food, but live only a few days without water. Under normal conditions, an adult human turns over something like 2.5 litres of water a day, with more than a litre of that being metabolised from food and the remainder coming from what we drink.

Food, heat and air go together. Anyone who's ever been on a diet has counted calories—or, these days, often kilojoules (one calorie is equivalent to 4.171 kilojoules). A calorie is the amount of heat required to heat 1 gram of water 1 degree Celsius at a pressure of 1 atmosphere. How do we get from this image of a mad scientist with a tripod, beaker, thermometer and Bunsen burner to those calorie counts that appear as holy writ in diet books, or are displayed in smallest-size print (unless they are expensive 'health' products) on food packaging? The manufacturer may decide to take the easy option and use the 4–9–4 multiplier rule to estimate the calorie content: after doing the chemistry to determine the composition of the particular product, the grams of protein are multiplied by 4, fat by 9 and carbohydrate by 4. The more dramatic alternative is to make the actual measurement using a bomb calorimeter, a gadget that confines what are essentially small, controlled explosions. A very strong, sealed, steel vessel containing the dried test sample is pressurised with pure oxygen to ensure complete incineration, then placed in a water bath before electrical ignition. The key measurement is the increase in the temperature of the surrounding water.

A normal, moderately active, human adult needs to eat a balanced diet that isn't deficient in key nutrients and vitamins and supplies a minimum of 1500 to 2500 calories per day. The divergence reflects variation due to age, size and gender. It's way beyond the scope of this discussion to go into either the physiology of nutrition or the details of what constitutes a healthy, balanced diet, but that information is readily accessible from web-based sources, nutritionists, your doctor, and the better diet books. The 1500 to 2500 calorie range is, in fact, a 'diet book' type of estimate. Those of us who read the packaging before we buy don't need to thrust the item back onto the supermarket shelves if the numbers are unexpectedly and horrifyingly large—just divide by 4.2.

The average US citizen eats and drinks about 3300 calories per day, while Kellogg's Australia recommends about 2700 calories per day. Fewer than 2000 calories per day is considered to be below the poverty line for those doing hard physical work in developing countries. Starvation is the inevitable consequence when this drops to 1000–1200 calories per day. An accompanying lack of iron and vitamin A leads to anaemia and blindness, a risk in some Asian societies where rice is a dietary staple. Other well-known diseases that result from vitamin deficiencies are beriberi (B-1), scurvy (C), pellagra (B3) and pernicious anaemia (B12). Even relatively inactive dieters who consume fewer than 1200 calories per day may be advised to take a multivitamin pill.

Evidence from rat and mouse experiments indicates that very low calorie diets can contribute to longevity, but there are risks with such strategies and it is probably advisable for anyone embarking on this course to first take advice, then be monitored regularly by a qualified professional. While studies with mice and rats tell us a lot about how complex biological systems work, whether drugs are likely to be safe, and so forth, the results cannot always be extrapolated directly to humans. Though you might be able to think of someone you know who fits the bill in one way or another, we humans are not rodents.

Successful dieting does, of course, confer major bragging rights. As a rule, we don't diet quietly. Recently, many of us have been trying one or other of the currently fashionable, high-protein, low-carbohydrate diets. The 'South Beach Diet' is earning a lot of kudos. The thing to watch out for with the ultra low-carbohydrate/high-protein approach is, I think, evidence of muscle wasting. If that happens, it's probably time to back off. Many took it further and added the high fat of the two-fried-eggs-and-lots-of-bacon Atkins diet. Though my friends who started down this road were jubilant about the rapid weight loss most of the Atkins enthusiasts I know have not persisted. One contributing factor was that, when Dr Atkins sadly and suddenly dropped dead, there were all sorts of rumours about the parlous state of his arteries. Similar cardiovascular problems have been reported for other Atkins acolytes.

The further consequence of severe dietary regimes is that, as carbohydrate stores are rapidly depleted, the body switches to breaking down fat and mobilising fatty acids to produce the essential energy supplier, glucose, by a process called gluconeogenesis. That 'ketone' type of metabolism is more normal in cows and bulls, which produce a lot of volatile fatty acids in the vast, bacterial fermentation vat they carry around, the fourth stomach, or rumen. Most humans, though, tend to be a bit put off when their breath takes on the sweet, acetone smell characteristic of ketosis. Even in cows, an excess of acetone in the air they exhale is not a sign of good health.

Given all of the above, the pure hot air diet doesn't look like a goer. Surprisingly, though, something very like it has been tried. The odd and (for obvious reasons) self-limiting Breatharians believed that, with proper spiritual development, it isn't necessary to eat or drink anything. Once a sufficiently high level of 'being' is achieved, the enlightened individual can apparently survive by simply breathing ambient air! The few poor, devoted souls who were sucked in by this insane idea succumbed, as might be expected, within a fairly short time. Even more dreadful is the thought that, because they failed to achieve the superior degree of spirituality that allowed them to live on air alone, they may have expired in a state of guilt and despair.

Those equally deluded (but presumably non-compliant) characters who led these tragic individuals down this fatal road were duly tried and locked up for a few years. How is it that some can stand aside and watch as gullible acolytes go to a certain, useless death? Unfortunately, this seems an all-too-common behaviour pattern for those who achieve undue power over the fate of others. The Breatharians were also considered for one of the Darwin Awards. The somewhat unfeeling citation reads: 'Named in honour of Charles Darwin, the father of evolution, the Darwin Awards commemorate those who improve our gene pool by removing themselves from it'. Such is the extent of human oddity that the Breatharians did not make it into the final Darwin list for that year. Others who received these posthumous accolades were involved in activities that were both lethal and even more deluded.

Though the Breatharians took things too far, spirituality and dietary discipline often go together. A university friend was not looking forward to returning to Thailand where, for a time, he would be required to live the life of a Buddhist monk with his begging bowl. He subsequently did his time, and survived to prosper. The images of the Buddha, though, do not suggest that lifelong dietary restraint is a necessity for spirituality. The saints and prophets of the Abrahamic tradition fasted in the desert. John the Baptist lived on locusts and honey, a high-chitin high-sugar diet. A few closed Christian monastic orders, although not followers of St John's particular regime, still adhere to a very plain and limited diet. Some other monastics were known for living a little too well, especially in mediaeval times. Henry VIII reacted accordingly, though motivated more by acquiring these monks' wealth than by any desire to purify the religious life and reduce their girth.

The contemporary observations of Lent, Ramadan, Passover and so forth serve to focus the mind on the shared values of the particular religion. As most of the underlying beliefs are good and life enhancing, this seems to be a sound discipline. A little dieting is probably beneficial for many of us and, in any case, it's not hard to understand that real psychological benefits ensue from short-circuiting, at least for a time, the increasing self-obsession and narcissism that seems to go with high levels of material prosperity. Also, as the faithful process through the streets bearing saintly relics and images, such festivals serve to remind the broader public in our predominantly secular, western societies of both their national history and the particular holy occasion.

We happened to be in Madrid before Easter a few years back, and saw lots of people dressed up in what looked like purple Ku Klux Klan outfits. These were the 'Nazarenos', some of whom, the 'Penitentes', also carried heavy crosses. The visual effect was powerful, especially as we initially made the mistake (because of the mental KKK connection) of assuming that the tall, peaked headdress and gown outfits reflected some sort of celebration of that darker side of the Spanish religious tradition, the Holy Inquisition. That era when the faithful from both the Protestant and the Catholic camps were burning each other in different countries has always had a horrible fascination for me. What is depressing is that this type of insanity seems to have re-emerged. As 9/11 and subsequent events have shown us, some are still so deluded and blinded by religious and/or political zealotry that they find it acceptable to incinerate innocent human beings who are somehow different.

The required limitation on food intake in ritualised fasting is never at a level that threatens survival or, as these religions emerged in pastoral and agricultural societies, even the capacity to work. Also, the period of restraint generally ends with some form of celebration, even if it's as simple as hot cross buns at Easter. The clothing industry benefits from the sale of elaborate outfits, including Easter bonnets and dresses, while the providers of various types of specialised food products can do a good business.

Religious dietary prohibitions can also be seen to make good practical sense when viewed from an historical viewpoint that emphasises food safety. My early training in veterinary science at the University of Queensland led coincidentally to being registered as a meat inspector. The killing floor of a major abattoir is about as down to earth and distant from the 'hot air' end of the human experience as it is possible to be! Though it may seem horrible to contemporary sensitivities, just about anyone who lived in mediaeval and earlier societies would have been exposed regularly to the slaughter and butchering of large animals. A good steak doesn't start out as a cling-wrap encased piece of meat in a supermarket chiller cabinet. The recent incidence of fatal H5N1 bird flu in Thai children reflected that seven-year-olds in rural villages are given the job of plucking the chickens for the family pot. That type of exposure can induce perhaps too pragmatic and utilitarian a view of the human condition, but it does lead to a respect for basic realities and, strangely enough, for the nature of domestic animals and our long-term relationship with them.

Both the Islamic and the Judaic traditions regard the pig as unclean, while in traditional agricultural societies, pigs and people often live in close contact. The bacterial diseases transmitted from pigs to humans include the gastrointestinal infections caused by *Balantidium coli*, *Campylobacter jejuni* and some strains of *Escherichia coli* and *Salmonella typhimurium*; and the severe febrile conditions Brucellosis (*B. Suis*), Erysipelas (*E. rhusopathiae*) and Leptospirosis (particularly

L. interrogans). Immature forms of the pig roundworm (*Trichinella spiralis*) can, if eaten in uncooked pork, circulate through the body and become encapsulated in muscles where they cause permanent weakness. The migrating larval stages of the pig tapeworm (*Taenia solium*) lodge in various human tissues and develop into the bladder-like Cysticercus form. Depending on the particular anatomical site, the space-occupying Cysticerci can, by simple pressure, induce brain seizures, heart failure and even death. Severe respiratory infections caused by influenza A viruses spread from pigs to humans, as can the newly identified Nipah virus that was a prominent cause of encephalitis (brain inflammation) in a recent Malaysian outbreak.

Though all these problems are generally well controlled by vaccination, meat inspection and other public health measures, it's easy to see how some early disease-detective may have made the connection and in turn recommended that it is best for humans and pigs to lead very separate lives. Similarly, a bad oyster or two, or deaths caused by 'Red tide' algal toxins ingested following the consumption of one or other filter feeder, could have led to the rule against eating animals without backbones. Interestingly, when I've run my pragmatic 'food safety' view of religious gustatory and ritual practices by some of my more devout friends, a few have found it discomfiting and even blasphemous.

It is intriguing that some prefer the idea of a divine mandate, while others incline to the view that intelligent, effective, human responses to substantial problems became codified as religious rituals in cultures that had no under-standing of the underlying biological processes. I guess this is the point where science and faith inevitably divide. While the believer may see such food prohibitions as a means of practising self-denial and adherence to the mysteries and dictates of faith, the scientist is more likely to accept that, as we probe underlying realities and develop new insights and evidence, our thinking and behaviour will change accordingly.

This is, in a sense, a clash of different types of human disciplines, each of which can have its separate psychological value. Where any one individual falls on the spectrum between those two approaches to life may, perhaps, be to some extent a reflection of the different ways our nervous systems are wired. As Gilbert and Sullivan have it in *Iolanthe*, 'Every little boy and girl that's born into this world alive is either a little liberal or a little conservative.' We should see over the next few years whether combining new scientific approaches based in genomics and functional magnetic resonance imaging (MRI) will provide substantial evidence for a measure of inherited determinism in the way that the brain works. Coincidentally, maybe we'll identify the *French Women Don't Get Fat* gene.

One candidate we already know about is the hormone Leptin. Leptin is produced by adipose (fat) tissue, then travels via the blood to interact with cells in the hypothalamus region of the brain. The net consequence is that we get a 'not hungry' message. The relatively rare individuals who have a genetic abnormality (mutation) in the receptor for Leptin do become very fat. So far, though, nobody has worked out how to use Leptin to treat obesity. As is often the case in modern medicine, it is the surgeons who, using approaches like stomach stapling or stomach bypass, provide the greatest relief to those who are grossly overweight.

Obesity resulting from the lethal combination of food abundance and a sedentary lifestyle is one of the great health problems of our age. Excess body weight is, together with the increase in longevity due to medical advances ranging from vaccination to the control of cardiovascular disease, contributing to a global pandemic of type 2 diabetes mellitus. Type 2 diabetes is traditionally a problem that emerges with age, but it is now being seen increasingly in young people and even in children. Juveniles are normally afflicted by type 1 diabetes, a condition that results from destruction of the insulin-producing pancreatic islet cells, perhaps as a consequence of some as yet unknown toxin or virus infection. On the other hand, type 2 diabetes is associated with being very overweight, abnormal glucose tolerance (blood levels are too high), insulin resistance and defects in insulin production. This global type 2 diabetes/obesity pandemic is a matter of serious concern and is the focus of a great deal of activity on the public health and medical research fronts.

The disaster is at its worst in populations that have undergone a very sudden transition from traditional lifestyles, such as the Polynesians of the South Pacific. While gathering protein-rich foods (fish) or complex carbohydrate sources (coconuts, yams) required a fair amount of effort, 'Coca-colonisation' has exposed them to a readily accessible but low-quality western diet that is high in sugar and refined carbohydrates. Paul Theroux's travel book *Happy Isles of Oceania: Paddling the Pacific* has some very depressing observation on the social and dietary consequences of westernisation for these formerly self-sustaining societies. Rather than make the effort to go fishing, it's easier to live on canned Japanese tuna and frozen chickens from Arkansas. Add readily available white flour, candy bars, alcohol and sugary drinks and the combination is lethal. Many die before they reach their fiftieth year.

Much of the scientific evidence suggests that the Polynesians, and similarly afflicted indigenous communities in nations that were colonised from the seventeenth century onwards, have not had the time to adapt in the genetic sense to this new 'prosperity'. The Israeli sand rat (*Psammomys obesus*) is a thin, active, saltbush-eating desert dweller that, when confined and fed a 'western' diet, rapidly becomes obese and develops type 2 diabetes. The 'thrifty gene' hypothesis proposes that people (or sand rats) who lived successfully under conditions where the food supply is sporadic or marginal are what those involved in animal production describe as 'good converters'. That is, they conserve the excess energy from food as fat and glycogen (carbohydrate) stores that can then be mobilised in times of hardship. Today, the optimal situation in western societies is to be slim, stylish and a 'poor converter'—that is, to show signs of what those involved in animal agriculture describe as 'ill thrift'. Most supermodels would not, I suspect, appreciate being showcased as stellar examples of 'ill thrift'.

Way back before the time of agriculture (at least 10 000 years ago) dietary limitation followed by binging must have been the ever-present reality for our earlier, hunter–gatherer ancestors. A restricted intake of seeds, fish, small animals and the like resulted from the painstaking, modest efforts of the female gatherers. This was then supplemented in a big way when the flamboyant, male hunters succeeded in despatching a mammoth or a kangaroo that had to be eaten before the kill went bad. The men then relaxed for a while and focused on painting cave walls and practising ritual magic. In many respects we haven't changed much over the years, though the roles of the sexes are now sometimes reversed.

Perhaps this 'deep genetic history' explains why we tend to overeat when we're travelling. The hunter–gatherer bit of our brain insists, against all reason, that what's plonked in front of us on the fold-down airline tray table is the last food we may see for a week. The microwaved 'beef Wellington' or 'Atlantic salmon' becomes the mammoth. Also, if you want to access high-flying, dietary hot air, read the colourful accounts of the virtues of the (albeit remote) chef in some of the more elaborate airline menus. As with my adolescent excursion into *Diet Does It*, we become so bored on a long flight that we will read (or eat) anything.

Even if hunter–gatherers could read, the need for diet books would be utterly incomprehensible to them. The same is likely to be true for the 800 million or so contemporary humans who don't get enough to eat each day. Severe malnutrition undoubtedly contributes to poor immune responses and enhanced disease susceptibility. In turn, parasite problems, particularly heavy roundworm (*Ascaris lumbricoides*) infestations, are a major cause of both calorie and iron deficiency in African children. Such conditions can be treated cheaply and easily, but the problem is part of a vicious cycle driven by poverty and degraded social infrastructures. The fifty least developed countries have poor roads, few services, literacy rates below 20 per cent and a GDP of less than US$800 per head of population. Unrestricted population growth and the HIV/AIDS catastrophe that takes out key people like farmers and teachers further fuel this prescription for disaster. Part of the solution lies with local economists, religious leaders and politicians who have the sophistication, determination and leadership qualities to promote behavioural change and fight corruption, social degradation and exploitation in these poor countries. This is a tough and often dangerous road to follow, and the west needs to give such heroes all possible assistance.

The other half of the equation is the altruism and international leadership required to persuade those in the league of wealthy societies that it is essential to divert more funds for AID and development. Each and every one of us can make a contribution in this regard. Every international AID agency needs additional financial resources. We can all ask questions of politicians and political parties, and vote accordingly in national elections. There is no simplistic right–left divide. The problems are complex and multi-factorial. Apart from increasing the quantity and quality of the locally available food supply, a major need in the developing world is to fund the health clinics that provide both preventive and emergency medical support, particularly for women and children. Some magnificent efforts in the area of disease control are drawing on private resources, particularly the Bill and Melinda Gates Foundation and the Merck Pharmaceutical Company, which continues to donate the drug Ivermectin to cure River Blindness, the disease that results when blackfly vectors (various *Simulium* species) carry the embryonic microfilarial form of the tiny worm *Onchocerca volvulus* to the human eye.

The rich nations could do a great deal more, especially by giving up a little so that poor people can help themselves. A persistent diet of hot air and distant promises is of no value to anyone, but this is often the only result of, for instance, the GO8 and G20 negotiations that seek to dismantle the trade barriers and agricultural subsidies that discriminate so cruelly against those in the developing world. Though the motivation may not be totally altruistic, Australia has been on the side of the angels in trying to change this situation.

Many of the best efforts of international aid agencies can be summarised by the old Chinese proverb: 'Give a man a fish and you feed him for a day. Teach a man to fish and you feed him for a lifetime'. That presupposes that there are any fish to catch. The parable of the loaves and the fishes told in all four of the Christian New Testament Gospels also deals with an acute food shortage and one way of handling the problem. Most of us, though, can't rely on a handy miracle when a rumbling stomach tells us that it's time to be fed. Travelling the highways and byways of the Mississippi delta, a likely lunch or dinner choice is to stop for southern barbecue, particularly barbecued pork ribs. If you're eating a low-fat diet, or you've seen the movie *Fried Green Tomatoes* and are reminded of the very remote possibility that you could unknowingly be consuming the local villain, then you could choose instead to patronise one of the occasional fish restaurants along the road. Forget the 'low fat' of the baked scrod and the 'flame-seared' ahi (yellowfin) tuna that might grace the menu at

Boston's Legal Seafoods or Sydney's Doyle's at Circular Quay. Fish in the rural delta is likely to be either buffalo fish or catfish. Very little money buys an enormous meal of deep-fried catfish and vegetables, fried potatoes (French fries, or 'freedom' fries since the disagreement with 'old Europe' over Iraq), coleslaw and beans in barbecue sauce. Washed down with beer or that weak American coffee, it tastes great if you're hungry, though eating this way too often can be a problem.

In days gone by, the catfish would have come from the mighty Mississippi and its associated creeks, backwaters and bayous. The catfish are still there, and some folks use their days off to go fishing on that dangerous, fast-flowing river in little outboard-motor powered aluminium boats, or 'tinnies' in the Australian vernacular. That may be the closest that the river culture now gets to the imagined world of Jim, Huck Finn, and the King and the Duke in *The Adventures of Huckleberry Finn*. It's not hard to imagine Huck and Jim cooking catfish over an open fire as they float downriver on their raft.

The nineteen-year-old Abe Lincoln must have eaten catfish as he crewed an unpowered flatboat loaded with pork, flour and potatoes on a thousand-mile journey down the Mississippi to New Orleans. That experience would also have given the future president some insight into the lives of slaves like Mark Twain's Jim, which certainly influenced his later political trajectory. When they reached their destination, the wooden flatboats were broken up for firewood and the crew either walked home or rode the huffing and chuffing paddle steamers.

Now, all commercial catfish are farmed in the big ponds dotted across the landscape of rural Mississippi or Arkansas.

Each 4000-acre pond can contain as many as 5000 catfish. Fish farming is big business. The channel catfish (*Ictalurus punctatus*) is one of the better fish to farm, as it is largely vegetarian and does not need to be fed large amounts of fishmeal. Fishmeal is made from the unacceptable parts of fish that we normally eat and from 'scrap fish', aquatic life forms that are not used directly for human food. 'Scrap fish' cannot be in infinite supply, and taking large numbers must eventually have an effect on the food chains of acquatic species that we do find palatable.

Farmed catfish are in every US supermarket. They taste great when cooked in batter with plenty of fat and salt, but are otherwise fairly bland and need some type of sauce or seasoning to make them interesting. The fillets of ocean farmed Atlantic salmon displayed alongside them in the fish-counter chiller cabinets make for much better eating, though farmed salmon does not have the extraordinary taste, colour or texture of, say, the wild-caught king salmon or cohoe salmon of the American northwest. All salmon are carnivores and, when farmed, consume much larger amounts of fish-meal. The same is true for the farmed barramundi that we buy in Australia. Still, unless we can discipline ourselves to conserve and harvest the fish of the ocean in a sustainable way, farmed fish are likely to be the future of 'seafood' consumption for many of us.

The other fish that appeared suddenly on US restaurant menus in the 1980s and 1990s was the orange roughy (*Hoplostehtus atlanticus*). A long-lived, slowly reproducing deep-sea fish taken first in large numbers from the cold waters off the coast of New Zealand, it has a firm consistency and a very characteristic taste. The combination of improved deep-sea fishing technology and jumbo jets meant that there was an immediate global market for this new fish experience. Just like the local restaurant owners, we could buy large bags of frozen orange roughy fillets quite inexpensively at Sam's Discount Warehouse in Memphis. Now many of the orange roughy fisheries that stretch across the Southern Ocean have been greatly depleted and are showing no signs of recovery. The Australian government recently listed it as a threatened species. In the United States, orange roughy was still a menu item last time I looked, though much scarcer and found mainly in the more expensive seafood restaurants.

Fish are cold blooded (or poikilotherms), which means that, unlike homeotherms (us and other furry animals), they vary their body temperature. Under very cold conditions they produce antifreeze-like proteins that allow them to live in, for instance, the deep ocean. Many fish should also be able to tolerate the increased water temperatures that are likely to result from global warming, so long as this does not disrupt the food supply that depends ultimately on small organisms like the zooplankton and phytoplankton. Those that stay in the one place and feed from reef ecosystems may, however, be in trouble if the local flora and fauna cannot adapt. Warmer water may not be a problem for farmed fish and, in any case, the industry could move further away from the equator in the face of rapid environmental change. It does make sense that we should learn to farm our favourite fish species, and to manage those that can't be farmed with much greater care and sensitivity.

As with any domestic animal, when fish are farmed and fed artificially, the higher densities make them much more susceptible to disease. The Norwegian Veterinary College in Oslo, for instance, has a big infectious disease and pathology program that is servicing the salmon-farming industry. Visiting some years back, I heard for the first time about a whole area of very impressive science that, like much of the technology that makes modern life possible, operates pretty much below the radar of media (and thus public) awareness. Oregon State University also has a major effort focusing particularly on diseases of salmon. The University of Tasmania and the University of Queensland both have programs on fish diseases and, as fish farming is expanding rapidly in Australia, this is likely to be of increasing commercial interest.

Bill Clem at the University of Mississippi in Jackson heads a substantial research group working on catfish immunity. When catfish ponds get cold, the fish tend to shut down their immune systems. As a consequence, they can be particularly susceptible to infections with fungi that grow at low temperatures. Infection isn't just a problem for farmed aquatic species that happen to have fins, scales and backbones. Bob Day at the University of Melbourne is focusing on a virus that was first detected in cultured abalone but also kills abalone in the wild when released into the environment. Viruses can wipe out oyster and shrimp (prawn) populations, both of which are now extensively managed and harvested. The shrimp that you see listed on the menu in Melbourne, Dallas or the Cayman Islands could well come from Thailand.

Bony fish, like catfish and salmon, have an 'adaptive' immune system like ours that can be stimulated by vaccination to give protective immunity. Such vaccines are already well established in the industry and there are active research and development programs to deal with continuing problems. Administered either by injection or in the food supply, some of these vaccines have been in use for twenty years. However, the immune systems of molluscs and crustaceans, like abalone, oysters, prawns and crabs, may only be capable of rather broadly specific, 'innate' responses. We have no real understanding of how (or if) we could make vaccines for these species. All are being farmed in one way or another, so this is a subject that should now be of considerable interest to us. If vaccination is not possible, then the management system will have to emphasise isolation, tight infection control and, perhaps, molecular engineering to introduce resistant genes.

While a major aspect of our involvement with fish focuses on the 'fast-food' bucket of fried catfish, the gas barbecue or a bottle of chilled riesling and fine china plates in an up-market 'slow-food' restaurant, there are many other dimensions to the interaction between human beings and the life forms in the seas and rivers of our planet. The fishing experience may differ in many respects from Isaac Walton's account in *The Compleat Angler*, but the essential aesthetic is still the same. Though most of the trout in mountain streams will have grown from farmed fingerlings, that reality doesn't in any sense diminish the skill or enjoyment of the fly fisherman. What better way to detach from life's everyday problems than to concentrate on the eddies and ponds of a fast-flowing river, or to stand at night on a warm tropical beach holding a rod and line thrown into the swirling ocean?

Fishermen who practise 'catch and release' respect the environment and are a powerful force in the constant fight to maintain, or restore, pristine watercourses and seascapes in the face of short-term commercial interests. Awareness accompanied by a sense of deep emotional commitment is a major dynamic when it comes to preserving the natural environment. That's just one reason why all who have the physical capacity should turn off the television set at the weekend and take every opportunity to get out into the mountains or onto the beaches.

Our relationship with salt water is ancient. Evolution tells us that our vertebrate ancestors crawled onto land from the ocean. The month-old human foetus has folds (the pharyngeal pouches) that embryologists have long regarded as primitive gill slits. The religious creationists detest this 'ontogeny recapitulates phylogeny' (development repeats evolution) gill slit idea and revile the argument as just so much duplicitous scientific hot air. Both sides emphasise (but from diametrically opposed viewpoints) that there is something 'fishy' when it comes to this piscatorial interpretation of mammalian development! (Isn't English a marvellous language? The use of 'fishy' to describe something that's fake, or dubious, evidently comes from the fact that our noses only detect the odour of fish oils when the process of decomposition is well under way.) Also, any cook or fisherman knows that fish are even more slippery than some politicians and fundamentalist preachers. The saying that 'fish rot from the head down' may also refer to presidents, prime ministers and CEOs, but where did we get 'crabby' (bad tempered)? Does it relate to the crabs of the ocean, or to the discomfort caused by 'crab lice' (*Pediculosis pubis*) infestations? Shrimp for 'small' is obvious, while being described as a 'prawn' in Australia is equivalent to being called a 'nebish' in Yiddish. A great saying from my Queensland childhood was 'don't come the raw prawn with me', meaning don't try to make a fool of me. 'Fishy' and 'hot air' are almost interchangeable terms when it comes to public discourse.

Snorkelling or diving on tropical reefs among the myriad brightly coloured fish and other life forms gives an incredible sense of wonder. For those who don't want to get wet there are glass-bottomed boats, underwater viewing facilities, minisubmarines and at least one marvellous, three-dimensional IMAX movie. Watching fish in a tank encourages contemplation and a sense of peace, and it is a magical experience to walk through a modern aquarium and suddenly come face to face with a shark or a grouper.

It will be an incalculable loss if ocean warming proceeds with such rapidity that it outstrips any capacity for adaptive evolution and results in the death of corals and the degradation of reef ecosystems. We should be working now on the science that might allow us to reproduce these environments at higher latitudes. There is no certainty that corals and the other, essentially local, life forms associated with them could make this geographic transition without our help. Even worse, if some of the predictions about increased acidification of the oceans are correct, it may become necessary to develop artificially controlled ocean 'theme parks' so that those who come after us can experience the magnificence of the coral reefs that we take for granted. What an immense tragedy it will be if we come to that! No amount of deliberate duplicity, or rewriting of history, will protect the 'legacy' of those leaders and generations who do not make the utmost effort to avoid such a catastrophe.

Becks and *Bleak House*

Preternatural combustion

The idea that sleeping human beings can suddenly burst into flame and be reduced to ash without exposure to some extrinsic source of fire or extreme heat sounds pretty suspect to most contemporary ears, but 'preternatural combustion', as this phenomenon was called, was taken quite seriously in the nineteenth century. It's discussed at length in the only book we have from Bill Stephens, Penny's father: a leather-bound, fifth edition of *Becks' Medical Jurisprudence*. Was it one of his undergraduate texts? He was a 1928 graduate of the University of Melbourne School of Medicine where I currently spend most of my working year. Even in those days, Medicine moved a lot faster than the Law.

Our copy of *Becks* was published in 1836. In the section titled 'Persons found dead', Theodric Romeyn Beck MD and John E Beck MD take eight pages to consider 'Of persons found burnt to death'. While in some cases the reasons for suffering such a horrific fate was very much in line with what happens today, the *Becks* devoted much of their account to the now unfamiliar preternatural combustion. In their words:

It is stated in the Transactions of the Copenhagen Society, that, in 1692, a woman of the lower classes, who for three years had used spirituous liquors to such excess that she took no other nourishment, having sat down one evening on a straw chair to sleep, was consumed in the night time, so that next morning no part of her was found but the skull, and the extreme joints of her fingers; all the rest of her body was reduced to ashes.

Spontaneous self-combustion wasn't restricted to the heavily imbibing lower classes, though that perception would have suited the emerging temperance movements of the nineteenth century.

The Countess Cornelia Bandi of Cesena, in Italy, aged sixty-two, and in good health, was accustomed to bathe all her body in camphorated spirits of wine. One evening, having experienced a sort of drowsiness, she retired to bed and her maid remained with her till she fell asleep ... the next morning ... at a distance of four feet from the bed was a heap of ashes, in which the legs and arms alone were untouched; between the legs lay the head.

The *Becks* cite the authoritative Le Cat, who described how the alcoholic wife of the Sieur Millet, at Reims, suffered a similar fate. 'The judges formed an opinion that he had conspired with his servant to destroy the wife, and he was condemned to death. On appeal to a higher court, however, this decree was reversed, and it was pronounced a case of human combustion.'

So far as modern science is concerned, human beings do not burn without the aid of some form of accelerant such as gasoline, napalm, or flaming bed-coverings lit by a fallen cigarette. In order to convert a human torso to ash in the way described by the Becks, a crematorium oven is generally run at 900 to 1000 degrees Celsius. It is doubtful that the Sieur Millet would fare too well with a twenty-first century court of appeal, though preternatural combustion does have a permanent place in literature.

Charles Dickens used this fire from within to dispose of the gin-soaked rag and bone dealer Mr Krooke in *Bleak House*. Discovered by the tyro lawyer Mr Guppy and his friend Mr Weevle, all that's left of the unfortunate Krooke is 'a smouldering suffocating vapour in the room, and a dark greasy coating on the walls and ceilings ... the cinder of a small broken log of wood sprinkled with white ashes, or is it coal?'

Bleak House was published in twenty monthly parts between March 1852 and September 1853, being pretty much the equivalent of a high-class soap opera of the day, the type of thing that Americans who watch PBS TV are used to seeing Alastair Cooke and more recently Russell Baker 'introduce' on 'Masterpiece Theatre'. The six-part video version of *Bleak House* aired in the 2005 season of *Masterpiece Theatre*. Viewing it in Australia on ABC TV without the benefit of an 'introducer' (not necessary in the old Imperial countries), we saw the permanently inebriated, reclining Krooke experiencing a greater than usual sense of inner warmth before he meets his incendiary end. That's one way to write someone out of the plot so that the action can move on. Dickens also uses the more conventional (for the time) devices of consumption and 'weakness of the chest' to kill off his redundant characters. The soap opera equivalent in these days of improved respiratory medicine is the car accident or a brain tumour.

Dickens attracted some criticism from the rationalists for the method he used to eliminate Krooke. In response, his 1853 preface cites the historian Le Cat and quotes the case of 'the Countess Cornelia de Bandi Cesenate'. He also seems to describe the trial, then acquittal, of the Sieur Millet. The preface finishes with: 'the recorded opinions and experiences of distinguished medical professors, French, English and Scotch, in more modern days; contenting myself with observing that I shall not abandon the facts until there shall have been a considerable Spontaneous Combustion of the testimony on which human occurrences are usually received'. Nobody writes quite like that any more, though the second- or third-hand accounts he cites hardly rate as evidence for a scientist, or even a modern defence lawyer.

As a research biologist who works with experiments and newly generated data sets, I found it kind of fun to be associated with what might be a very small finding in literary scholarship. It was, in fact, Penny's insight, not mine. Dickens is surely quoting *Becks' Medical Jurisprudence* as his authority when discussing the fate that he assigns to

Mr Krooke. For a moment, we both felt just a little of the smouldering fire that burns in the mind of the library-oriented scholar as he or she pores through dusty tomes and exhumes reams of previously ignored, long-forgotten correspondence. It was in order to put just such a bunch of ribbon-bound letters into circulation that Dickens found it necessary to remove Krooke. Then we looked on the web (http://www.anomalyinfo.com) and found: 'SHC [spontaneous human combustion] an Anomalies Study'. It was very clear that the *Bleak House/Becks* association has already been amply explored.

Garth Haslam, the author of *Anomalies*, further informs us that Le Cat's first name was Nicholas, Madame Millet was Nicole, her husband was the proprietor of the Lion d'Or (used later as a book title by Sebastian Faulks), and the Appeals Court concluded that she died 'by a visitation of God'. Furthermore, it seems that many of the Becks' accounts of preternatural combustion were derived from Jonas Dupont's 1763 book *De Incendiis Corporis Humani Spontaneis*.

This brief experience is sufficient to convince me that lighting new intellectual fires by perusing the published word is far from easy; so, as someone who's driven by curiosity, I think it's best to stay with experimental science. My guess is that it would be pretty hard to get a research grant to work on preternatural combustion and there is, in any case, no relevant animal model. People have described alcoholic rats but, though they can develop testicular atrophy, there is no record that they burst into flames. In any case, it is inconceivable that any responsible university ethics committee would approve a strategy that involves spontaneous combustion of the subjects. Unless we see further natural occurrences, which seem to be conspicuously lacking in this era of forensics and scientific medicine, preternatural combustion will have to remain beyond the reach of the scientific method, in the realm of mystery, myth and literature.

Political Hot Air

On buffoons and the commentariat

Unless we are so subdued in character that we conform to the biblical injunction 'But let your communication be; Yea, yea; Nay, nay; for whatsoever is more than these cometh of evil' (Matthew, 5:37), a certain amount of verbal hot air is pretty much normal for the human condition. A common conversational device is to overstate a case in order to see what response comes back. Ambit claims that advance inflated positions tend to mark the beginning phase of any industrial or peace negotiation. Also, whether politician, poet or physicist, we are all fallible beings who can hold and express sincere opinions that are later shown to be just plain wrong. Our brilliantly presented, magnificently formulated, incisive insights may turn out in retrospect to be so much hot air. Depending on the way you view the world, you may feel exactly that way about some of what's written in this book, particularly where I've touched on the global warming issue. Time will tell whether many of us are blowing too hot about the long-term, building problem of hot air.

The technique of confusing issues by using a mass of words in a formal political speech now seems almost Dickensian. In *Bleak House*, we meet two forms of historic, human hot air. The spontaneous corporeal conflagration that reduced poor Mr Krooke to little more than a greasy substance and a bad smell was already discussed in the previous chapter. Though we are not actually told that another, more minor character, Mr Gusher, is a politician, he seems to command the heights of verbosity that qualify him for admission to the lower registers of nineteenth or early twentieth century political life. In Chapter VIII, entitled 'Covering a multitude of sins', Mrs Pardiggle, a lady of 'commanding deportment', asks:

> You know Mr Gusher? ... He is a very fervid and commanding speaker—full of fire. Stationed in a wagon on this lawn, now, which from the shape of the land, is naturally adapted to a public meeting, he would improve almost any occasion you could mention for hours and hours.

Today we don't have time to listen to any politician for hours and hours, or even for minutes. If we have moments to spare from our over-busy schedules, we have a pressing need to watch telecasts of four different types of football, baseball, tennis, lawn bowls, hockey, billiards, racing, sailing, basketball, cycling, synchronised swimming, ping-pong, poker, cricket, the Olympic Games, the Commonwealth Games, the Pacific Games and the Whatever Games. Failure to do so may mean that we exclude ourselves from the common dialogue and deny our origins as tribal beings. It's much better to be part of a tribe that prefers to watch men kicking footballs to men blasting away with AK47s or Armalites. If we're not into sport, then there are various money, idol, survival, desperate housewives, desperately dancing, travel and cooking programs. Among my diversions are those TV detective 'soaps' where people of all conditions, shapes, sizes and accents are wantonly murdered and lined up on post-mortem tables. That probably helps to defuse any murderous tendencies I may have.

In this contemporary iteration of the visual 'bread and circuses' culture that has provided fertile ground for political manipulators since (at least) the time of the Roman Empire, we're just too intellectually and emotionally exhausted to pay attention to some pompous, self-serving political monologue, even if it does address issues of substance. That type of deliberate distraction from underlying realities can end very badly, as we saw in the inexorable progression from the care-fully engineered, spectacular triviality of Louis XIV's court at Versailles to the storming of the Bastille and the series of incisive meetings between many members of the French ruling class (including Louis XVI and Marie Antoinette) and Madame La Guillotine. As Dickens relates in *A Tale of Two Cities*, the latter were colourful and entertaining events for some, like the industrious Madame Defarge, though not for those who had previously enjoyed the high life. When one segment of society grabs too much at the expense of others and refuses to address real problems there is, in the end analysis, always a reckoning.

To return to the mellifluous Mr Gusher, we do get to meet him, accompanied by Mrs Pardiggle and Mr Quale, in Chapter XV of *Bleak House*.

> Mr Gusher, being a flabby gentleman with a moist surface, and eyes so much too small for his moon of a face that they seemed to have been originally made for someone else, was not at first sight prepossessing; yet he was scarcely seated before Mr. Quale asked Ada and me, not inaudibly, whether he was not a great creature—which he certainly was, flabbily speaking; though Mr. Quale meant in intellectual beauty—and whether we were not struck by his massive configuration of brow.

At least by the definition of the former Labor leader Mark Latham that 'Politics is showbiz for ugly people', Mr Gusher does seem to qualify for life as an elected representative in a contemporary parliamentary democracy. Where he might do better financially, though, is to use his talent for hot air by becoming a political pundit in the print or broadcast media where, again, good looks are not a pre-requisite. Whatever his underlying personality and beliefs, Mr Gusher would be able to fit in somewhere. The members of this well-paid, professional commentariat range in character and presentation from the reflective and reasonable to some who blow back and forth with the winds of public opinion and the perceived wishes of those who dispense patronage, through to a few hectoring extremists who are clearly captured by some deep, personal psychopathology. Come to think of it, that's also a pretty good description of the human spectrum that we elect to govern us. Gusher would no doubt have been amply equipped to function in either capacity.

One good thing that is happening, given the current concentration of ownership of much of the conventional media, is the growing prominence of various well-resourced, intellectually driven blogs. Screening the on-line websites of quality, daily newspapers throughout the world also allows us to retain perspective and to remain informed. Keeping the internet free and open has to be a major priority if we are to continue benefiting from the diversity of ideas, approach, debate and political compromise that has so promoted human well-being over the past few centuries. In the long run, the web may be our best means of maintaining the broad communication, disclosure, information exchange and responsive dialogue that makes democracy work and protects both us and our world. We must not allow this medium to be stolen from us by established, old-style communications companies and their political surrogates.

In recent times, the hot air put out by organised political parties has been largely distilled into short, professionally controlled, super-heated bursts. Television broadcast time is expensive, and the nature of the visual media does not encourage a sustained focus on any issue unless it can be made visually engaging. In US politics particularly, wedge slogans like, 'gay marriage', 'right to life', 'pro-choice' and 'big-spending liberals' reduce complex, many-layered issues to sound bites. The only truth that matters in advertising is an increase in market share, whether the market is for some widget or the vote in a democratic election.

Embedded too solidly in my memory bank is an emphatic 'Salad shooter! salad shooter!' part of a sales jingle for a gadget that, as I recall, jetted sliced vegetables into a bowl. Who could live happily without such an important piece of culinary equipment? The TV election spots highlighting 'good old X' from 'the party that cares' are just as trivial, though less memorable. Depending on our own political orientation, we sometimes turn those messages around in our own minds to come up with something that's the opposite of what the advertiser intended. I always heard Newt Gingrich's 'Contract with America' as 'Contract on Americans' which, I think, pretty much turned out to be the case.

The truth of the matter is that the human brain seems best equipped to retain short, catchy snippets that fit with our particular view of the world. Positive memories of politicians often relate to succinct, profound statements that resonate through time. When Abraham Lincoln gave his Gettysburg address, delivered on that grievous, bloody battlefield during the course of a Civil War that was yet to be resolved, some of those who were present to hear the speech were disappointed with its brevity. Less than two minutes and 269 words long, the President's speech is totally memorable and ends with: 'and that government of the people, by the people, for the people shall not perish from the earth'. We care nothing for Secretary of State Edward Everett's two-hour oration given during the course of the same occasion.

Then there's the great deal of good sense that came out of the mouth of Teddy Roosevelt, including: 'Character, in the long run, is the decisive factor in the life of an individual and of nations alike', or 'If you could kick the person in the pants responsible for most of your trouble, you wouldn't sit for a month'. A message that resonates for many is Franklin D Roosevelt's line in his inaugural address: 'the only thing we have to fear is fear itself', a call to action that seems particularly relevant in the politics of today, though it was intended to introduce the 'new deal' that ended the 1930s Depression. My respect for politicians is greatly increased when I hear them repeat President Harry S Truman's 'The buck stops here' on those inevitable occasions when something goes badly wrong. One prime minister who did speak truth was Malcolm Fraser when he said: 'Life wasn't meant to be easy'—although, as he was perceived as a patrician, that wasn't terribly popular with much of the electorate.

President Dwight D Eisenhower was frighteningly prescient when he said that 'we must guard against the acquisition of unwarranted influence, whether sought or unsought, by the military-industrial complex. The potential for the disastrous rise of misplaced power exists and will persist'. Surely everyone knows the statement from John F Kennedy's 1961 inaugural: 'And so, my fellow Americans: ask not what your country can do for you—ask what you can do for your country.' Less prominence is given to the sentence that followed: 'My fellow citizens of the world: ask not what America will do for you, but what together we can do for the freedom of man.'

The US leadership of the 'free world' that was firmly in place in the era of presidents Eisenhower and Kennedy is much less assured as I finish this in mid-2007. The ill-judged military adventure in Iraq has done immense damage to American credibility. The culture of political denial and the attempts at deliberate suppression of the science related to global warming that characterised the Bush 43 White House until very recently have, perhaps for the first time since World War II, passed the torch on a major international issue to Europe. My guess is that this will be repaired once the disastrous Bush presidency grinds to its eagerly anticipated end. Under Republican governor Arnold Schwarzenegger, California, which (with 36-plus million people) is both as big as one of the larger European nations and a site of major technological innovation, is clearly committed to rapid change. Many other US states are now taking up the challenge.

A major factor driving a sea change in the orientation of 'conservative' politicians and commentators is that the 'king-maker' in US, British and Australian politics, News Ltd's Rupert Murdoch, made it very clear during the latter part of 2006 that he has switched from being a global warming sceptic to recognising that this is a major issue that must be addressed. The Murdoch media empire includes the immensely influential (in the United States) Fox Television Network and many tabloid and 'quality' newspapers, including *The Times* and *The Australian*. Within weeks of Mr Murdoch's confession there seemed to be a significant shift in both attitude and rhetoric on the part of the conservative Australian prime minister John Howard. We can only hope that the Murdoch press will now hold the feet of politicians to the fire if they fail to promote carbon trading, emission controls and renewable energy research in ways that go beyond the realm of token actions, promises for the future and deliberate lies.

The word in 2007 among political acquaintances on both the right and the left of US politics was that the next Republican nominee for the presidency will be drawn from the ranks of those who are committed to decisive action to reduce greenhouse gas emissions. That was certainly the 'take home' message when I heard Senator John McCain, a leading contender for the 2008 Republican nomination, speak in Washington. The dynamic among senior Democrats, spear-headed by former vice-president Al Gore, is such that it is inconceivable that a future Democratic president will fail to address this issue. For some time, British Prime Minister Tony Blair led the charge for change in Europe and it seems certain that his successor, Gordon Brown, is set to follow suit.

Because of the nature of what they do, and because the law of unintended consequences can play out in diverse ways over a substantial time interval when it comes to international affairs, the actions of presidents and prime ministers are best left to the measured judgement of history. Their policies often come back to haunt them. One certainty is that they will wish to be remembered positively for the substance of their enduring achievements, not for their failures or for the heat of their more outrageous rhetoric. The capacity to distil the consequences of various political decisions and actions in the calm light of retrospective analysis is what we expect from professional historians. With the global warming issue, though specialists will relentlessly sift and dissect the data on who said or did what, such scholarly activity may not be needed to tell us about consequences. All we may require to make that evaluation is a thermometer, an altimeter to measure where we sit above sea level, and a water gauge.

Flying the Concorde

Turbofans and afterburners

The needle-nosed Anglo–French Concorde must be far and away the most aesthetically pleasing jet airliner ever built. I flew in it once about twenty years ago, from London's Heathrow to Washington's Dulles airport, after spending a very grounded week at Nairobi's International Laboratory for Research in Animal Diseases. Like any agriculture-oriented operation, ILRAD was a pretty pragmatic and down-to-earth place. I'd been mixing all week with scientists who may be intellectually elegant but are, at least in the sartorial sense, generally a rather dilapidated and down-market bunch. After passing through security in London, it was quite a contrast to find myself in the executive Concorde lounge. It certainly wasn't my scene.

My no-doubt selective recollection of the lounge was that it was largely populated by stocky, deeply tanned, middle-aged guys wearing designer suits and little, slip-on shoes with gold buckles. Grecian formula was the vogue, and many were adorned by the presence of much younger and very glamorous women who probably weren't their daughters. After Dulles, the plane went on to Miami. No ectomorphs here: they were certainly neither 'my people' nor men that it would be good to have as enemies.

Boarding was quick, as the plane wasn't full and the Concorde carried only 100 passengers anyway. The cigar-tube cabin was narrow and constricting after the wide-body that I'd flown up on from Kenya, while the fittings betrayed the 1960s origins of the plane. There were no movie screens, with the only display being a big gauge at the front of the cabin that showed air speed in Mach values.

Though in a front bulkhead seat, I did not have the privilege of being in the Concorde cockpit. Once, in those simpler days before 11 September 2001, we sat with the flight crew on a Qantas 747 en route from Sydney to Auckland. What impressed me was that the plane did not move forward until the engine exhaust was hot enough. The gas-burning Rolls-Royce turbofans had to get up to temperatures in excess of a thousand degrees Fahrenheit before take-off. I'd never thought about it but, whenever we fly in a jumbo we are, in effect, being propelled by the hot air expanded in four big and very sophisticated fossil-fuel burning 'primus' stoves. Like doctors and nurses, the pilots monitor the 'good health' and efficiency of their charge by taking its temperature!

That's very simplistic, of course, because jet engines also use multiple turbines to increase thrust by compressing the incoming air. This process is so efficient that turbofans don't even need to achieve full combustion. Supersonic jets like the Concorde achieve their great speeds by using afterburners to consume the residual oxygen. Though the afterburners that pump in more fuel at the end of the normal exhaust cycle were only used to get the plane up to maximum velocity, they did make the Concorde a 'gas hog'.

'The take off will be a little sporting!' warned the British Airways captain, and the Concorde bumped along with rapidly increasing speed till we suddenly shot up into the air. Flying sub-sonic over England and Ireland wasn't much different from any other trip, but then we left the Emerald Isle behind and started over the Atlantic. The clouds below began to move much more quickly than I'd ever experienced. There was no doubt that we were travelling very fast indeed. The Mach gauge climbed progressively through 1.6, 1.8 to over 2.0—twice the speed of sound. A Boeing 747 travels only at sub-sonic speeds.

Concorde passengers had no physical experience that they'd been propelled through the sound barrier, but the sonic boom experienced by those on the ground was a major factor in the ultimate commercial failure of this spectacular aeroplane. What each of us hears as a single boom from a fast-flying jet is, in fact, a continuing bellow generated by the plane's turbulent 'shock wave' as it breaks through and exceeds the normal 'speed limit' of the surrounding air molecules. People don't enjoy having their ears assaulted or their windows broken on a regular basis, so Concorde was never allowed to fly super-sonic over land. Together with its high fuel consumption and relatively short range, this meant that the only good run for it was across the Atlantic. The Pacific was too wide, and it offered no advantage for the land-based routes of Europe and Asia.

Though I was oblivious to the bellow, what I remember most about the Concorde trip was that it was far from relaxing. There wasn't that sense of the mindless disconnect that results if you've been travelling or working all day, then sink into a comfortable seat that someone else has paid for on a late, 12- to 14-hour flight across the Pacific. Apart from the short flying time, the speed of the Concorde imposed a unique sense of urgency, of physicality. Though there isn't enough oxygen for us to breathe normally and survive at twenty kilometres above the earth's surface, battering through this thin air at Mach 2 results in friction, heat and constant vibration. The cabin lining felt hot when I reached my hand out to touch the side of the plane. Those diffuse molecules of oxygen, nitrogen and even a little carbon dioxide were resisting our passage and making their presence felt. There was nothing passive about this flying experience, this violation of the atmosphere.

Concorde effectively halved the trans-Atlantic trip time, taking only about three and a half hours. After being served a nice lunch, we were on the ground again and I was soon on a standard commercial jet en route to my middle-America destination. The Concorde trip was the ultimate plane flight for me, though I never felt the need to repeat the experience. Still, for the professionals who flew those jets and those who designed and built them, there must be a sense of lost possibilities and nostalgia for a venture that was in the heroic tradition of human ambition and technology.

Ultimately though, the Concorde was one adventure that proved just too costly to sustain. It ended in tragedy when a bit of debris on a Paris tarmac fragmented a tyre, which, in turn, threw up a chunk of rubber during that 'sporting take-off'. The resultant penetration and fuel-tank leak to those hot engines in the wings then led to the horribly spectacular pictures of the rapidly ascending aeroplane trailing streams of fire, and soon crashing back to the ground, that we all remember. Combining oxygen and jet-fuel with the temperature of a furnace is fine so long as it's confined and controlled within the chamber of a jet turbine, but life is fragile and death is the consequence when that fire escapes. Whether a newer, but less environmentally damaging, version of the Concorde will emerge phoenix-like at some future time is anybody's guess.

It took less than one human life span to go from Orville and Wilbur Wright's short flight over the sands of Kitty Hawk to the Red Baron's tiny triplane to the extraordinary adventure of the supersonic Concorde. Step back another 120 years, and we are in the time of Louis XVI, Marie Antoinette and the first hot air and hydrogen balloons. Neither the Wright brothers nor our unfortunate Royal couple could have even imagined the elegance, speed and beauty of the Concorde.

Heating the Planet

From charcoal to wind farms

Who could be in doubt that the human condition has been transformed over the past 300 years? Most would agree that the change has been for the better. Electricity not only lights but also heats and cools our houses, moving the source of combustion and environmental pollution to remote power plants. We don't splutter from respiratory congestion and our eyes don't water as we cook dinner or read in bed. Anyone can fly on commercial flights and circumnavigate the planet in less than seventy-two hours, providing they are crazy enough to do so and can afford it. Jules Verne's *Around the World in 80 Days* is as historic as Homer's *Odyssey*. We love our cars because they give us extraordinary freedom of movement and, at least for some, define our personality and status in society.

For many, friends and relatives are dispersed around the globe. We circulate jokes and family photographs, submit manuscripts, download music and conduct casual conversations by e-mail. The same is true for minor business dealings when, sometimes to our chagrin, we find ourselves speaking by telephone to an amiable and (by western standards) grossly underpaid person in Bangladesh. We no longer look up a print dictionary or encyclopaedia as a matter of first resort, but immediately use a search engine to check out Wikipedia and other independent sites. Powering our computers, communications networks and the plethora of associated electronic devices takes a lot of electricity and the demand is constantly increasing, though it should mean that shakers and movers who are on the international circuit don't have to jump on jet planes quite so often.

These profound changes in the way we view life and operate in the world are a direct consequence of the industrial revolution and the revolution in electronics and communications that followed. In truth, though, we could regard them as the first and second phases of the 'energy revolution'—the transformation that resulted from burning fossil fuels. The extraordinary developments in manufacturing and transport that saw the continued expansion of British power and wealth throughout the 1800s were coal-fired, while oil consumption drove the incredible rise of the United States during the twentieth century.

In effect, accessing these sources of stored, concentrated energy allowed the focus on evidence, reason and the systematic acquisition of knowledge that characterised the European enlightenment to be translated for material advancement. The early entrepreneurs/inventors of the age of coal, iron and steel were followed by the emergence of an ever more innovative industrial sector facilitated by a high-end, science-based culture in major universities, dedicated research organisations and company-based research and development programs. What we've shown, particularly over the past century, is that our capacity for creative thought and action is such that, if we can access enough energy, it's possible to overcome just about any problem that has a solution based in engineering and technology.

Perhaps the greatest achievements of the energy/science/technology revolution are the extraordinary improvements in health, and thus human potential, over the past 150 years. The availability of clean water, vaccines and antibiotics has dramatically reduced childhood mortality rates in all but the poorest and most dysfunctional societies. Heart drugs and advances in surgery, dentistry and so forth have allowed more people to enjoy longer, relatively pain-free lives than at any time in the 120 000-plus year history of *Homo sapiens sapiens.* Contrast the few bad teeth remaining in the head of the aging Louis XIV with the gleaming dentition of today's more modest senior citizens. Dentistry seems such a mundane business, but think what tooth pain, or not being able to chew because of lack of teeth, does to your disposition, clarity of thought and nutritional status.

Although that's all to the good, these marvellous advances have also created a major problem. While the global population is thought to have increased from about 300 million to one billion in the 1800-year interval from the birth of Christ to the beginning of the industrial revolution, the next 200-plus years saw the number of people on this small planet multiply a further six-fold to 6.4 billion. Think about it:

there are now twenty times more hot air-generating, CO_2-producing human beings than there were when the rule books of the great Abrahamic religions were written. Nobody drove around in Hummers in those dim, distant and poorly recorded days. Not only that, we keep vastly greater numbers of methane-producing cows and sheep to provide meat, milk, wool, leather and (in the poorest countries) traction. Methane (natural gas, CH_4) is considered to be forty times more detrimental than CO_2 when it comes to the greenhouse effect, even though, if collected and burned, it is (as discussed earlier) the cleanest of the carbon-containing fuels.

Obviously, while children are both our greatest delight and the future of our species, the current rate of human population growth can't continue. Most of the increase in numbers over recent years has been in the predominantly agrarian societies of the developing world. The continuing demand for more arable land, together with the economic opportunities afforded by logging to provide charcoal (for cooking and heating), timber and paper, has led to the massive destruction of tropical rainforests that function as global CO_2 sinks. Soil quality is often poor, and tillage exacerbates the process of environmental degradation and species loss. One positive sign is that some of the advanced countries (including Australia) are now recognising the problem and are providing substantial aid to help reverse the process of deforestation. Overall, if we are to deal effectively with the greenhouse gas problem, we need to think very seriously about how to achieve a more equitable international distribution of wealth applied to achieving positive ends.

We might expect to see smaller families in what are currently the poorer nations of the planet if the much-vaunted globalisation of economic opportunity indeed translates into broadly beneficial outcomes. The 'farmer's insurance' of a large family is no longer needed once there is access to a comprehensive social security system. The inverse correlation between increasing prosperity and population size is very obvious. People in the advanced democracies are having fewer children, or choosing to remain childless, with the effect in Japan and Italy being sufficient to raise concerns about how these rapidly aging societies will function in the future.

While we may eventually stabilise (or even decrease) the number of us, just having fewer people on the planet will not solve the greenhouse gas problem. Greater wealth is associated with the increased consumption of energy and manufactured goods. Most of the developing countries don't even register on the CO_2 emission charts that are being shown by everyone from environmentalists to captains of industry seeking to establish their 'green' credentials. As might be expected, the nation at the top of both the material prosperity and carbon production leagues is the United States. However, China has been industrialising very rapidly and, with four times as many people has now surpassed the USA as the world's greatest carbon emitter.

Apart from CO_2, winds from the west also carry a plethora of small, particulate products of combustion from the smog-shrouded Chinese cities and industrial parks to the neighbouring countries of Korea and Japan. India also has a prominent place in that toxic aerosol league, though the formerly awful situation in New Delhi has been significantly improved by banning buses, cars and auto-rickshaws that are more than fifteen years old and switching many new vehicles to natural gas. The brown haze over Asia weakens both the Asian and Indian monsoons and, drifting south, is modifying rainfall patterns in northern Australia. The problem is not confined to the developing world. Air quality in Houston, Texas—the spiritual capital of the oil industry—is appalling and is also thought to be suppressing rainfall. Texas abuts Louisiana, with its concentration of major oil refineries. The nearby states of Mississippi, Tennessee and Alabama recently experienced the worst droughts on record.

Even if we set aside the greenhouse effect associated with gases like CO_2 and methane in the upper atmosphere, the massive air pollution that is blighting so many beautiful parts of the world should lead us to the inescapable conclusion that we have to change the way we do things. Rubbing the flask that released the twin genies of coal and oil to provide the magic of the energy revolution has come with a price that we must now pay or, by diligence and imagination, seek to mitigate. Many people are depressed by the thought that we are caught in an unholy trinity of environmental degradation, rapid loss of plant and animal species and inexorable greenhouse gas-induced climate change. As recent, authoritative analyses like the British Stern Review and the quinquennial statement from the International Panel for Climate Change (IPCC) emphasise, that doesn't have to be true, providing we start the remedial process now.

What we must look to is the third phase of the energy revolution, the next step that allows us to tap sources of renewable energy derived directly from the sun, the winds, the ocean, geothermal heat buried deep within the earth and (if we can ever get it contained) nuclear fusion as replacements for the fossil hydrocarbons. Solar energy can be harnessed directly or by growing plants to produce biofuels, though these still come with a carbon component. We probably do need to exploit clean-burning coal technology that utilises gasification, hydrogen production and carbon sequestration as an interim measure. At least for a time, it seems inevitable that nuclear fission has a part to play, particularly in the colder countries of the northern hemisphere.

But we should leave the funding of coal and nuclear research to the private sector that has profited (or will profit) so handsomely from the sale of these products, and use our unencumbered (by established industry interests) tax dollars to support genuine innovation and product development in the renewables sector. It's a very depressing fact that global investment in research aimed at developing sustainable solutions to the energy problem has been falling since the oil shock of the late 1970s. The more we can fund smart people to explore a diversity of both short- and long-term strategies, the better off we, and our successors, are likely to be.

Much of what has to be done initially is at the level of public policy, both locally and globally. The national political process in both Australia and the United States is particularly susceptible to becoming the intellectual and moral captive of geriatric, established industries that, despite the 'spin' they may put out, have little intention of changing their ways while profits remain high. That, by the way, does not describe all energy companies, particularly some of the electricity producers that are very open to the idea of using any feasible alternatives to coal and oil. The same division is true for the automobile manufacturers, some of whom can't seem to imagine how it might be possible (in the United States) to achieve an average of thirty-five miles per gallon (14.8 kilometres per litre) by 2020, while others (particularly in Asia and Europe) have already reached that goal for mid-sized cars and four-wheel drive vehicles that use gas/electric hybrid or turbo diesel engines.

The first thing each and every one of us needs to do is make the effort to educate ourselves and reach our own conclusions. I'm as much a layman as anyone else when it comes to global warming. The only thing that my research field of viral immunity has in common with this area is complexity, and the associated possibility that outcomes may vary in unpredicted and even catastrophic ways. Time and time again over a long research career, I've been jarred out of my complacency by the joint operation of the 'law of unintended consequences' and Murphy's Law (anything that can go wrong will go wrong). That's OK when we're talking about a study with laboratory mice, but not so good when the big 'experiment' involves all the life forms on this planet. Some of the current climate change scenarios are very scary, and even scientists who do know a lot about this stuff point to the risk of unexpected, dramatic 'tipping points' that could go very badly from our point of view.

Trying to engage with this field, I've read a few popular books (listed in the Bibliography) and viewed Al Gore's PowerPoint presentation *An Inconvenient Truth*. The former US vice-president's Oscar-winning documentary is also available in book form and is a good place to start for anyone who wants to see the basic arguments presented in a clear and easily understood way. Some of the consequences he illustrates may be overstated. If the current IPCC predictions are correct, Gore's rather dramatic scenario for sea level rise will not happen in the twenty-first century. However, the IPCC considered that the published data sets for global ice melt were not sufficiently comprehensive to be incorporated in useful predictions, so they focused on the effect that simple warming would have on ocean volume. The expanded use of satellites equipped with radar altimeters is providing a lot more information on the thickness of the ice sheets, so it is likely that the 2012 IPCC report will deal much more directly with this issue.

A little less accessible to a broad audience than *An Inconvenient Truth* are the excellent 'News and Views' summaries in the top weeklies *Nature* and *Science*, which nearly all scientists read. Science is not monolithic and, though all fields are evidence-based, their approaches, language and technology vary greatly. As a consequence, scientists depend on well-written, clear reviews to learn what is happening in important areas outside their own narrow discipline. Of late, there is something on climate change in almost every issue of both journals. Also, because my research interests stretch to the area of global infectious disease, I speak from time to time at major forums that are addressing the problems of the developing world, and get to hear others who are discussing topics like water availability, drought, desertification, ocean acidification, the destruction of coral reefs, species loss and various other consequences related to ever increasing atmospheric CO_2 levels and global warming. Among the best talks I've heard in these types of meetings have been very direct, forward-looking analyses from executives in the energy and insurance industries.

Anthropogenic (man-made) global warming can hardly be avoided in a discussion of hot air, but I viewed the prospect of reviewing this area with considerable trepidation when I started writing some eighteen months ago. What has happened in the interim is that the sceptics (deniers in some people's book) have essentially been routed, and even the most 'conservative' politicians are now engaging (or pretending to engage) in a positive and forward-looking way. At a recent summit, I heard a leading financial expert who can hardly be considered to come from the 'greenie' or 'left' side of the political spectrum state: 'The science of climate change is done, we now need to move on to develop solutions.' He's not quite right: no science is ever completely 'done' and we have a great deal to learn about these complex processes.

As we're all aware, the issue of climate change and how to deal with it is receiving ever more attention in newspapers and on television and radio. Much of that discussion fits with what I've seen earlier in the scientific press. Other op-ed pieces and very readable accounts provide new insights that I hadn't even thought about. There is, though, the occasional article, commentary or television program that looks to be way off the mark. It's important to realise that science, like politics, religion or any other major field of human activity, has its fringe dwellers and monochromatic zealots. Such people can provide fertile sources for sensationalist reporters, who cherry-pick from all over the place to support a contro-versial position. They claim to be doing a public service by presenting a balanced view, even if the balance of opinion among specialists in the area is 99 to 1 per cent, yet they simply end up confusing people.

Seizing on this or that piece of isolated information is not how science works, especially when it comes to complex systems. Particularly in the area of climate change, it's essen-tial to integrate data and insights coming from a diversity of areas of expertise and sophisticated monitoring systems. The role of the IPCC is to reach agreed positions that make sense of the whole, providing a basis for useful recommendations to policy makers and information for all of us. Constituted under the auspices of the World Meteorological Organization (WMO) and the United Nations Environment Program (UNEP), the IPCC draws on the voluntary efforts of thou-sands of scientists in 192 participating countries. With this breadth of involvement it's inevitably a cumbersome and slow-moving mechanism. Also, the review process relies on already published information that, while it does have the advantage of first being scrutinised by some sort of peer review process, will not be right up to the minute.

The fourth in the quinquennial series of IPCC summary reports started appearing from February 2007 and the mate-rial is readily available on the web. 'Overall, the report concludes that global average temperature will rise between 1.1°C and 6.4°C by 2100, and that it is "very likely" (90 per cent certainty) that human activities and emissions are causing global warming.' The hope is that, by acting now, we can keep that down to no more than 2 degrees Celsius, but my guess is that this will be very difficult to achieve. Others suggest the truly terrifying possibility that mean temperatures could increase by up to 11 degrees, a change that would have truly disastrous consequences when occurring over such a short interval.

Nothing in this area is set in stone and, because of the complexity inherent in variables like cloud cover, ocean currents and tidal systems, the potentially negative effect of increasing temperature on the capacity of the oceans to act as a CO_2 sink, and so forth, the predictions that are being made now concerning consequences over the next century will inevitably be refined as time goes by. I'm not going to attempt to summarise how these various parameters might interact, as that could be done much better by any of the experts who participate in the IPCC process. Looking from the viewpoint of a science-trained layman, I'll just comment on a few situations and possibilities that pique my interest.

Apart from the drought that we've been living through in Melbourne, Australia, where I now spend much of my year, what engages me most is the whole issue of ice melt, glacier loss and so on. Growing up in the sub-tropics, I didn't even see snow until I was more than twenty years old. Having lived after that in colder parts of the world, nothing gave me a greater sense of peace and wellbeing than watching snow fall, breathing the cold, crisp air and listening to the silence in that too-brief interval between when the roads are impassable and when the ploughs come out. The thought of losing that fills me with great sadness.

While the senses and felicitous memories fuel emotions and the human spirit, what drives good science is intellectual insight, accurate measurement and an absolute respect for rigorously acquired information. Some of the clearest data sets available in the whole climate change area are for melting glaciers. The total surface of the glaciers worldwide has decreased more than half since the end of the nineteenth century and they are retreating with ever increasing rapidity. Satellite mapping analysis indicates that the Arctic, Greenland and West Antarctic ice sheets are contracting more quickly than had been suggested by earlier predictions based on modelling approaches. If all the Greenland ice goes, ocean levels could rise by up to seven metres, and, though nobody is suggesting that this is imminent, it is a far cry from the maximum of 0.6 metres (till 2100) suggested by the IPCC. The last interglacial period of 125 000 years ago was characterised by increases of four to six metres.

Unlike Greenland and the Antarctic, the Arctic pole has no land beneath. It's possible that, by 2050, the Arctic could be ice-free in summer, making the 1845 search for the elusive North West Passage that took the two ships and the lives of Sir John Franklin and his crew seem a remote curiosity. Estimates suggest that sea levels could rise by more than fifty metres if all the Antarctic ice also went. We can't allow that to happen. Significant evidence of erosion is being seen for the vulnerable West Antarctic ice shelf and, while Antarctic ice melt is currently thought to be causing a miniscule rise in sea level, it's clear that we need to do whatever is possible to stop this process from accelerating.

Decreased snow accumulation is also of great concern when it comes to the future availability of fresh water. A good deal of the water supply for the south-western United

States comes from snow melt in the Rockies. Santa Fe, Arizona, has now been in drought for years. Snow melt in the Himalayas feeds a number of major river systems, including the Mekong and the Yangtze, which irrigate much of South-East Asia. Hundreds of millions of people would be affected. In the long term we may start to see fresh water flowing inland rather than out to sea, the source being large, coastal desalination plants!

Much of the energy for the new $2 billion-plus desalination plant near Wonthaggi (in south-eastern Victoria) will be provided by a large wind farm. Palmerston North, New Zealand, now powers itself by more than 150 of those big wind turbines sited on the surrounding hills. Conventional electricity generators provide backup when needed, but the wind turbines compensate by feeding excess capacity into the grid when, as is generally the case in this city of 30 000 homes and 75 000 people, the breezes blow with sufficient intensity. In many places, anyone who has solar panels installed on the roof of their house can also send unused energy back along the wires to their provider and be compensated accordingly. One of the debates in the electricity industry concerns the extent to which generating capacity might be centralised or distributed, with many small solar 'plants' supplementing a new generation of clean and green power stations.

Palmerston North has a substantial population of educated and aware people, being the home of Massey University of Manawatu and the country's prestigious Crown Research Institutes. Overall, New Zealanders tend to be tough and realistic people, who dealt very successfully with an extremely grim set of economic realities when their tied markets for primary produce were massively compromised by Britain's sudden entry into the EU. Another university town, Vaxjo, in Sweden, is tackling the problem of climate change with even greater determination. By 2010, Vaxjo's 80 000 people intend to have their 'carbon signature' down to 50 per cent of what it was before they began their various lines of action. In the United States and Australia, the current debate is whether that will be possible by 2050.

Al Gore is advocating a 90 per cent reduction in that time frame and the people of Vaxjo are, I believe, telling us that this is not totally unrealistic. Many more of Vaxjo's citizens now walk a lot or ride bicycles and are thus at much less risk of developing type 2 diabetes. You might say that sort of action is easy in such a well-organised and rational society, but think what it's like to be on a bike in the middle of a Scandinavian winter. The basic lesson is that individual and community action is enormously important, though isolated initiatives by relatively small numbers of thoughtful and aware people will obviously not be enough.

Sweden has a population of just over nine million, so the Swedes could argue: 'Whatever we do is irrelevant, so why do anything?' But that's not valid. The world is in desperate need of strong leadership and decisive action on this issue. In the first instance, both have to come from the wealthy, technologically advanced western countries. How can we expect the nations of the developing world to address this problem if the rest of us do nothing? Apart from improved technology, we need meaningful carbon trading schemes, carbon taxes and, where governments continue to act irresponsibly, carbon tariffs. The latter might be remitted back to the poorer countries in the former of clean technology to minimise carbon emissions. Achieving an appropriate cost for carbon now is essential if we are to unleash the full power of the capitalist system in a search for sustainable solutions.

An incredible amount of creative thought and energy is going towards developing technological fixes. A recent issue of *Science* summarised a broad spectrum of innovative approaches, ranging from sliced solar cells to microbial engineering. The various carbon capture and storage (CCS) strategies that are being widely discussed generally depend on pumping CO_2 from new, 'integrated gasification combined cycle (IGCC)' coal-fired plants back into space deep within the earth that was previously occupied by 'harvested' oil and natural gas. A stern warning to coal-producing states is the statement by Vassily Kougionas, a European Commission officer in charge of clean coal initiatives: 'Without CCS there is no point in continuing with fossil fuels.' Australia, with its large coal reserves and export markets, has to make the IGCC process work.

A very different CCS approach is to process biomass to make both charcoal and fuel, then return the char back to the earth as a means of enriching the soil. Soil has an immense capacity to store carbon, which it releases back to the atmosphere following heavy tillage—yet another good reason for avoiding further, heavy-duty land clearing. One example is that of Danny Day, who runs the 'not for-profit social-purpose enterprise' Eprida. Based in Athens, Georgia, he builds contraptions for use on individual farms. For instance, Emma Marris relates that, 'Day's pilot plant processes 10–25 kilograms of Georgia peanut hulls and pine pellets every hour. From 100 kilograms of biomass, the group gets 46 kilograms of carbon—half as char—and around 5 kilograms of hydrogen, enough to go 500 kilometres in a hydrogen-fuel-cell car.' The great thing about this type of strategy is that, overall, more carbon is removed from the atmosphere than is put back. Also, local, low-technology approaches like this can be used anywhere.

How extraordinary that, over the brief interval of the past 300 years, we vulnerable, thin-skinned humans have managed to change the climate of our only home, this green and pleasant earth. I find that realisation very sobering, and believe that many who span the full spectrum of professional expertise and political loyalties are coming to much the same realisation. This isn't part of a left–right confrontation, or an issue that should divide religious believers and non-believers. Rather, it concerns the deep conflict between realists and fantasists that is a central feature of contemporary society. No amount of spin, deception or verbal hot air will make the global warming problem go away.

Exciting new ideas about how we might deal with climate change are emerging constantly. This light history of hot air has no pretensions to being a serious treatise on global warming and what to do about it. What's worth reading? The informed critics who reviewed Tim Flannery's *The Weather Makers* for *Nature* and *Science* were very positive, though they did have some minor reservations. I thought that Ian Lowe's brief account, *A Big Fix: Radical solutions for*

Australia's environmental crisis, was excellent—and equally relevant to the situation in many other countries.

We all need to look seriously and come to our own conclusions. As you do that, though, don't leave out of the equation the consideration of related factors like environmental degradation, the depletion of plant and animal species, and air pollution. If we take the trouble to look, each of us can see those dynamics at work in our own backyard. These are part of the same problem that relates directly to how we currently behave and live in the world.

What I can say is that, after assessing the evidence to the extent that my limited time and specific scientific expertise permit, I have the sense that anthropogenic climate change is the biggest, and potentially most dangerous, challenge that humanity has ever faced. I'm also convinced that, if we are to ameliorate, and eventually reverse, what's happening, we must start to act decisively now. We can do it, but not if we lie to ourselves and prevaricate.

As so succinctly stated by the physicist Richard Feynman: 'Reality must take precedence over public relations, for nature cannot be fooled.' That should be the guiding principle as we vote, exercise our purchasing power and organise our lives and communities to ensure a continued, bright future for humanity.

Notes and Selected References

Floating in Air

Ambrose, S 1997. *Undaunted Courage: Meriwether Lewis, Thomas Jefferson and the opening of the American West*. Simon and Schuster, New York.

Atwood, A 2003. *Burke's Soldier*. Penguin Books, Melbourne.

Bonyhady, T 1991. *Burke and Wills: From Melbourne to myth*. David Ell Press, Balmain.

Carey, P 2000. *True History of the Kelly Gang*. University of Queensland Press, St Lucia.

Lewis, H, & Killip, N 1996. *South Parkville*. The Parkville Association, Melbourne.

Schama, S 1989. *Citizens: A chronicle of the French Revolution*. Alfred A Knopf, New York.

Yule, P (ed.) 2004. *Carlton: A history*. Melbourne University Press, Melbourne.

Alphabet Soup

Anissimov, M 1999. *Primo Levi: Tragedy of an optimist.* The Overlook Press, Peter Mayer Publishers, New York.

Cornwell, J 2003. *Hitler's scientists: Science, war and the Devil's pact.* Viking Penguin, New York.

Levi, P 1975. *The Periodic Table.* English translation, Schocken Books, New York, 1984.

Snow, CP 1959. *The Two Cultures.* Cambridge University Press, Mass.

Life, Gas and Hydrocarbons

Isaacson, W 2003. *Benjamin Franklin: An American life.* Simon and Schuster, New York.

Soaring with Eagles

Bullock, A 1992. *Hitler and Stalin, Parallel lives.* Alfred A Knopf, New York.

Burleigh, M 2000. *The Third Reich: A new history.* Hill and Wang, New York.

Fullilove, M, & Flutter, C 2004. *Diaspora: The world-wide web of Australians.* The Lowy Institute, Sydney.

Nolte, E 1969. *Three Faces of Fascism.* Mentor. The New American Library, New York.

Pizzey, G, & Knight, K 2002. *The Field Guide to the Birds of Australia.* 7th edn, P Menkhorst (ed.), Harper Collins, Sydney.

Shirer, WL 1960. *The Rise and Fall of the Third Reich: A history of Nazi Germany.* Pan Books, London.

Smith, JS 1990. *Patenting the Sun: Polio and the Salk vaccine.* William Morrow and Company, New York.

Vonnegut, K Jr 1991. *Slaughterhouse-Five*. Dell Publishing, New York.

Wiesel, E 1972. *Night Trilogy: Night, dawn, the accident*. The Noonday Press, New York.

Burnt by the Sun

Page 55, 'The high numbers of immigrants' Poorly pigmented humans weren't the only European imports to suffer the consequence of exposure to high UV levels in northern Australia. Eye cancer (ocular squamous carcinoma) is a major problem in the decorative, white-faced poll Hereford cattle that, until fairly recently, figured prominently in the beef industry of sub-tropical Australia. As many as 24 per cent of the condemnations (for use as human food) in Herefords coming to slaughter at more than five years of age were due to eye cancer. As a consequence, producers turned to the Droughtmaster and Santa Gertrudis breeds that mix the genetic background of the highly productive *Bos taurus* with the more disease and sun-resistant *Bos indicus*. Others have bred Herefords with pigmented eyelids.

Clark, RR 1997. *The Mayne Inheritance*. University of Queensland Press, St Lucia.

Keneally, T 1998. *The Great Shame: A story of the Irish in the old world and the new*. Random House, Sydney.

Iron Horses and Balladeers

Foote, S 1958–74. *The Civil War: A narrative trilogy. Fort Sumter to Perryville, Fredericksburg to Meridian, Red River to Appomatox*. Vintage Books, New York.

Hearth and Home

Cannon, C et al. 1995. *Wildlife of Greater Brisbane.* Queensland Museum, Brisbane.

Crossland, BJ 1966. *Victorian Edinburgh.* Wayfair Publishers Ltd, Letchworth.

Malouf, D 1985. *12 Edmondstone Street.* Penguin, Melbourne.

McCall Smith, A 2005. *Espresso Tales.* Polygon, Edinburgh.

Pryde, GS 1962. *A New History of Scotland.* Vol. II, *From 1603 to the present day.* Thomas Nelson and Sons, London and Edinburgh.

——1965. *A New History of Scotland.* Vol. I, *From the earliest times to 1603.* 2nd edn, Thomas Nelson and Sons, London and Edinburgh.

Viser, BB 1998. *Central Gardens: Stories of a neighbourhood.* Paulsen Printing, Memphis.

Youngson, AJ 1966. *The Making of Classical Edinburgh.* Edinburgh University Press, Edinburgh.

The Iceman Cometh

Blainey, G 1966. *The Tyranny of Distance.* Sun Books, Melbourne.

Mullis, K 1998. *Dancing Naked in the Mind Field.* Pantheon Books, New York.

O'Neill, E 1999. *The Iceman Cometh.* Vintage, New York.

Turner Hospital, J 2003. *North of Nowhere, South of Loss.* University of Queensland Press, St Lucia.

Night Lights

Courtenay, B 1998. *The Potato Factory.* Penguin, Melbourne.

Dickens, C 1841. *The Old Curiosity Shop.* http://www.bibliomania.com/0/0/19/41/frameset.html

——1859. *A Tale of Two Cities*. Signet Classics, New York.

Johnson, EJ, & Russell, RD Jr 1990. *Memphis: An architectural guide*. University of Tennessee Press, Knoxville.

Keneally, T 1998. *The Great shame: A story of the Irish in the old world and the new*. Random House, Sydney.

Melville, H 1851. *Moby Dick*. Bantam Classics, New York.

Pepys, S 1660–9. *The Complete Diary*. http://www.pepys.info/

Shackleton, E 1920. *South: The* Endurance *expedition*. Signet Publishing, New York.

Imagining the Red Baron

'Vigilant' 1933. *German War Birds*. John Hamilton Ltd, London.

Barker, P 1991. *Regeneration*. Plume, New York.

Brenchley, F, & Brenchley, E 2002. *Stoker's Submarine*. Harper Collins, Sydney.

Brittain, V 1933. *Testament of Youth*. Wideview Books, USA, 1980.

Carlyon, L 2001. *Gallipoli*. Macmillan, Sydney.

——2006. *The Great War*. Pan Macmillan, Sydney.

Faulks, S 1993. *Birdsong: A novel of love and war*. Vintage, New York.

Field, SE 1943. *Singapore Tragedy*. Angus & Robertson, Sydney.

Gibbons, F 1932. *The Red Knight of Germany*. Cassell & Co., London.

Gibbons, LG (Mitchell, JL) 1932–4. *A Scots Quair: A trilogy of novels. Sunset Song, Cloud Howe, Grey Granite*. Schocken Books, New York, 1977.

Johns, WE 1932–86. The 97 Biggles Books are listed on the WE Johns and Biggles Website: http://www.collectingbooks and magazines.com/captain.html

Johns, WE 1941–1950. The 11 Worrals books are listed on Roger Harris's WE Johns website http://www.wejohns.com/Worrals

Keegan, J 2000. *The First World War*. Vintage, New York.

Keneally, T 1975. *Gossip from the Forest*. Hodder & Stoughton, London.

Koske, J 1987. *Ships That Shaped Australia*. Angus & Robertson, Sydney.

Lawrence, DH 1923. *Kangaroo*. Penguin, Melbourne, 1963.

Lee, L 1991. *Moment of War: A memoir of the Spanish Civil War*. The New Press, New York.

Liddell-Hart, BH 1932. *Foch, The Man of Orleans*. Greenwood Publishing, 1980.

Roth, P 2004. *The Plot Against America*. Jonathan Cape, London.

Speer, A 1976. *Spandau: The secret diaries*. Pocket Books, New York.

Thompson, P 2005. *The Battle for Singapore: The true story of the greatest catastrophe of WW2*. Portrait, London.

Treloar, JL (ed.) 1933. Australian Chivalry. Reproduction in colour and duo-tone of official war paintings. Australian War Memorial, Canberra.

Beacons

Durrell, L 1957–60. *The Alexandria Quartet: Justine, Balthazar, Mountolive, Clea*. Penguin, New York, 1991.

Mahfouz, N 1956–7. *The Cairo Trilogy: Palace Walk, Palace of Desire, Sugar St*. Everyman Library 2001, New York.

Stick, D 1952. *Graveyard of the Atlantic: Shipwrecks of the North Carolina coast*. The University of North Carolina Press, Chapel Hill.

——1958. *The Outer Banks of North Carolina: 1854–1958.* The University of North Carolina Press, Chapel Hill.

Tall Ships, Black Gangs, 'Bully' Wars

Goodwin, DK 1994. *No Ordinary Time: Franklin & Eleanor Roosevelt: The home front in WW2.* Touchstone Books, New York.

Miller, N 1992. *Theodore Roosevelt: A Life.* Quill, William Morrow, New York.

Phillips, K 2007. *American Theocracy.* Penguin Books, New York.

Rickover, HG 1976. *How the Battleship* Maine *was Destroyed.* United States Naval Institute, Annapolis, 1995.

Firefighters

Vonnegut, K Jr 1965. *God Bless You Mr Rosewater.* Dial Press Trade Paperback, 1998.

The Hot Air Diet

Agaston, A 2004. *The South Beach Diet: Good fats good carbs guide.* Rodale Publishing, New York.

Atkins, RC 1995. *Dr Atkins' New Diet Cookbook.* M Evans, New York.

Banting, W 1869. *Letter on Corpulence.* http://www.lowcarb.ca/corpulence/index.html, and Cosimo On-Demand Publishing.

Diamond, J 2003. 'The double puzzle of diabetes'. *Nature* 423, 599–602. Nature Publishing Group.

Guiliano, M 2004. *French Women Don't Get Fat: The secret of eating for pleasure.* Knopf, New York.

Hauser, G 1952. *Diet Does It: Incorporating the Gayelord Hauser cookbook.* Faber & Faber, London.

Mayson, IM 1861. *Mrs. Beeton's Book of Household Management.*
http://www.mrsbeeton.com

Theroux, P 1992. *The Happy Isles of Oceania: Paddling the Pacific.*
GP Putnam's Sons, New York.

Twain, M 1884. *The Adventures of Huckleberry Finn: Tom Sawyer's
companion.* Electronic edition by dell@wiretap.spies.com.
Released to the public July 1993.

Vernon, T 1983. *Fat Man on a Bicycle.* Michael Joseph, Penguin
Group, New York.

Watson, D 2005. *Death Sentence: The decay of public language.*
Knopf, Random House, Sydney.

Becks and *Bleak House*

Becks, TM, & Becks, JB 1836. *Beck's Medical Jurisprudence.*
5th edn, Longman, London.

Dickens, C 1853. *Bleak House.* Wordsworth Classics,
Hertfordshire.

Faulks, S 1989. *The Girl at the Lion d'Or.* Vintage, New York
1999.

Political Hot Air

Latham, M 2005. *The Latham Diaries.* Melbourne University
Publishing, Melbourne.

McCullough, D 1992. *Truman.* Simon and Schuster, New York.

Schultze, J (ed.) 2004. 'Webs of Power', *Griffith Review.* ABC
Books, Sydney.

The Huffington Post. Blog. http://www.huffingtonpost.com

Thomas, H 1999. *Front Row at the White House: My life and times.*
Lisa Drew/Scribner, New York.

Williams, P 1997. *The Victory.* Allen & Unwin, Sydney.

Heating the Planet

The following is an eclectic list selected mainly from my reading over the past 18 months or so. It is not meant to be in any sense comprehensive.

Alley, R et al 2007. *Climate Change 2007: The Physical Science Basis.* Summary for Policymakers. Draft of Working Group 1 contribution to the fourth assessment report on the Intergovernmental Panel on Climate Change (IPCC). Paris, 12 February 2007 http://www.ipcc.ch

Bindschlader, R 2006. 'Hitting the ice sheets where it hurts'. *Science* 311 (5768) 1720–21.

Bohannon, J 2007. 'IPCC report lays out options for taming greenhouse gases'. *Science* 316, 812–14.

Brumfile, G 2006. 'Academy affirms hockey-stick graph'. *Science* 441 (7097) 1032–3.

Clery, D 2006. 'Climate change demands action, says UK report'. *Science* 311 (5761), 592.

Deutsch, C. 'Sunny side up. Companies find a risk-free way to adopt solar power'. *The New York Times* Business Day, 21 October 2006, B1 and B8.

Doherty, PC 2003. 'Science, Society and the Challenge for the Future'. 7th U Thant Lecture, United Nations University, Tokyo http://www.unu.edu/uthant_lectures/index.htm

——2006. 'Beyond greed', in 'Hot Air: How nigh's the end'. *Griffith Review*, ABC Books, Sydney.

Dowdeswell, JA 2006. 'The Greenland Ice Sheet and global sea level rise'. *Science* 311 (5763), 963–4.

Flannery, T 2005. *The Weather Makers: The history and future impact of climate change.* The Text Publishing Company, Melbourne.

Giles, G, Schiermier, Q, Hopkin, M, Odling-Smee, L, Marris, E, Butler, D, and Corbyn, Z 2007. 'Climate Change/Business Section': From words to action; Behind the scenes; What we don't know about Climate change; Data keep flooding in; Climate sceptics switch focus to economics; What price a cooler future? Carbon copies; Market watch; Super savers; Experimenting with efficiency. *Nature* 445, 578–91.

Gore, A 2006. *An Inconvenient Truth: The planetary emergency of global warming and what we can do about it.* Melcher Media, New York. The DVD of *An Inconvenient Truth*, starring Al Gore, is directed by Davis Guggenheim.

Gullison, RE et al. 2006. 'Tropical forests and climate policy'. *Science* 316, 985–6.

Kennedy, D, & Hanson B 2006. Editorial: 'Ice and History'. *Science* 311 (5768) 1673.

Kerr, RA 2007. 'Pollutant hazes extend their climate-changing reach'. *Science* 315, 1217.

Lovelock, J 2006. *The Revenge of Gaia: Earth's climate crisis and the fate of humanity.* Basic Books, New York.

Lowe, I 2005. *A Big Fix: Radical solutions for Australia's environmental crisis.* Black Inc., Melbourne.

Luthcke, SB, Zwally, HJ, Abdalati, W, Rowlands, DD, Ray, RD, Nerem, RS, Lemoine, FG, McCarthy, JJ, & Chinn, DS 2006. 'Recent Greenland ice mass loss by drainage system from satellite gravity observations'. *Science* 314 (5803), 1286–9.

Marris, E 2006. 'Black is the new green'. *Nature* 442 (7103), 624–6.

Revkin, AC. 'Budgets falling in race to fight global warming. Big ideas, hard reality. Experts foresee flooding, drought and strife over energy'. *The New York Times*, 30 October 2006, A1 and A14.

Schiermeir, Q 2006. 'Putting the carbon back. The hundred billion tonne challenge'. *Nature* 442 (7103), 620–3.

——2006. 'The costs of global warming'. *Nature* 439 (7075) 375.

Schrag, DP (2007). Sustainability and Energy Section: 'Energy for the long haul; A sustainable future if we pay up front; Steering a national lab into the light; Eureka moment puts sliced solar cells on track; How to make biofuels truly popular; Small thinking, electrified froth and the beauty of a fine mess; Wiring up Europe's coastline; Hydrogen economy? Let sunlight do the work; Threading the nuclear fuel cycle minefield; Photovoltaics in focus; Rethinking mother nature's choices; Former marine seeks a model EMPRESS; Catalyzing the emergence of a practical biorefinery; Don't forget the long-term fundamental research in energy; Towards cost-effective solar energy use; Challenges in engineering microbes for biofuels production; Biomass recalcitrance-bioengineering plants and enzymes for biofuels production; Ethanol for a sustainable energy future; Renewable energy sources and the realities of setting an energy agenda; Preparing to capture carbon'. *Science* (315), 781–813.

Schulze, ED, & Freibauer, A 2005. 'Carbon unlocked from soils'. *Nature* (7056), 205–6.

Stern Review of the Economics of Climate Change, HM Treasury 2006. http://www.hm-treasury.gov.uk/independent_reviews/stern_review_economics_climate_change/stern_review_report.cfm.

Velicogna, I, & Wahr, J 2006. 'Measurements of time-variable gravity show mass loss in Anatarctica'. *Science* 311 (5768), 1754–6.

Abbreviations, Terminology
and Conversions

AE2	World War I Australian 'E class' submarine built by Vickers Armstrong
AK47	Russian assault rifle (Kalashnikov) used widely in the disadvantaged world
Ar	argon
Armalite	family of US military rifles
ARP	air raid protection, a function of civilian wardens in World War II
Ag	silver, argentum
Au	gold, aurum
BCE	before 'current era' described by the Gregorian calendar
BE2	World War I 'Bleriot Experimental' biplane designed by Geoffrey de Havilland
Biggles	fictional World War I and World War II fighter pilot Major James Bigglesworth

C	carbon, also used to describe vitamin C, or L-ascorbic acid
C_2H_2	acetylene gas
C_2H_6	ethane gas
C_2H_6O	ethanol, the form of alcohol we drink
C_3H_8	propane gas
C_4H_{10}	butane gas
$C_6H_{12}O_6$	glucose
CaC_2	calcium carbide
Camel	British WW I biplane with a cowl (hump) covering the forward machine guns
CCS	carbon capture and storage of CO_2 from coal-fired power plants
CE	year of the 'current era' described by the Gregorian calendar
CERN	European organisation for nuclear research and particle physics
CH_3OH	methanol, or wood alcohol
CH_4	methane gas
Cl	chlorine
CO, CO_2	carbon monoxide, dioxide
Dewar	vented flask used to contain liquid nitrogen for the storage of biological samples
DNA	deoxyribose nucleic acid, the stuff of genes, the hereditary material
Fe	iron
Gigaton	a billion metric tons
H, H_2	hydrogen
H_2O	water
H_2S	hydrogen sulphide, 'rotten egg' gas

H_2SO_4	sulphuric acid
HCL	hydrochloric acid
He	helium
Hisso	Hispano Suiza V8 engine that powered the SE5A biplane
HMS, HMAS	Her (His) Majesty's (Australian) ship
IGCC	integrated gasification combined cycle, to concentrate CO_2 from burning coal
ILRAD	former International Laboratory for Research in Animal Diseases, now ILRI (International Livestock Research Institute Nairobi, Kenya)
IPCC	International Panel on Climate Change that monitors research on global warming
Jet A	kerosene-like fuel for jet aeroplanes
K	potassium
kJ	kilojoule, unit of heat equivalent to 4.171 calories
lb	pound weight in the Imperial system of measurement, $=0.453$ kg
Lipitor	brand name for Co-A reductase inhibitor used to treat high blood cholesterol
LNER	London North Eastern Railway
LPG	liquefied petroleum gas
Maxim	machine gun used by the British (Vickers) and the Germans (Spandau) in World War I
N, N_2	nitrogen
N_2O	nitrous oxide, laughing gas
Na	sodium
NaCl	table or sea salt
Nano	10^{-9}, general abbreviation for very small, e.g nanoscience, nanomachines

NH_3	ammonia
NH_4	ammonium
Nm	nanometre, 10^{-9} metres
O, O_2	oxygen
OSM	various meanings, but used here to denote a member of the Franciscan order
P	phosphorus
PABA	para-aminobenzoic acid, used for UV protection
PBS	the US Public Broadcasting Service
Pullman Porter	sleeping car attendant on US trains of the twentieth century
RAF	British Royal Air force
RMS	Royal Mail ship, or steamer
S	sulphur
SE5A	British World War I 'Scout Experimental 5' biplane fighter
SMS	Same as HMS, but German
Spermacetti	oily material recovered from the head of whales that was used in candles
SPQR	Senatus Populesque Romanus, the senate and the Roman people
SS	steamship
SS	Shutzstaffel, the elite and murderous Nazi 'blackshirts'
Tripe	World War I Fokker triplane (Dreidekker) flown by the Red Baron
TSS	twin screw steam, referring to a ship
turnback	steam tank engine that has no attached tender
Uhlans	Prussian cavalry

USS	United States (naval) ship
USSR	now defunct, Union of Soviet Socialist Republics
UV	light in the ultraviolet band of the spectrum
UVA	320–400 nm wavelength
UVB	290–320 nm wavelength
UVC	100–290 nm wavelength
V8, V12	internal combustion engine with two banks of 4 or 6 cylinders in a 'V'
WAAF	British women's auxiliary air force of World War II
Worrals	fictional World War II aviator Flying Officer Joan Worralson
WPA	US Works Progress Administration, created under Franklin Roosevelt to end the Depression
Zn	zinc
$ZnCl_2$	zinc chloride, a salt

Conversions

1 inch	2.54 centimetres
1 foot	0.31 metres
1 metre	3.28 feet
1 mile	1.6 kilometres
1 degree Celsius	multiply by $\frac{9}{5}$, then + 32 to give Fahrenheit
1 pound	0.45 kilogram
1 kilometre/hour	0.62 mile/hour

All dollars are in Australian dollars unless stated otherwise.

Acknowledgements

The broad scope of this book goes way beyond my—or, for that matter, anyone's—area of professional expertise, and I am greatly indebted to colleagues and friends who work in other disciplines for discussion and advice. In particular, Paul Zimmett, Graham Pearman, Frank Witherspoon, Peter Joubert, Michael Doherty and Ian Doherty helped clarify some complex issues for me. In addition, I am particularly grateful to Tony Klein, who read the text and has hopefully ensured that I have not made any major mistakes when dealing with issues that have their basis in the physical sciences and engineering.

Discussions with Mary Cunnane and Louise Adler helped define the project early on, while Mary also provided critical insight by reading the first drafts of the various chapters. Penny read the manuscript as it developed, and her corrections were invaluable. It was a pleasure to work with Foong Ling Kong, who edited and re-ordered the text. Important sources are referenced, and I also made extensive use of web-based material and summaries in the leading scientific periodicals, *Science* and *Nature*.

Index

acetylene, 127, 134
Aden, 191, 193, 194
AE2, 141
Africa: and diseases, 225–6; environment and wildlife of, 3–4; landscape of, 3; North Africa, 163–4; *see also* global warming
AID agencies and foundations, 226
air: and flight, 7, 13, 37–8, 42–3, 170; particulate matter in, 7, 9, 260–1; and plant species, 8–9; political hot air, 241–2, 244–7, 249; rhetorical hot air, 42, 43, 44, 46, 209, 227; *see also* balloons
Alexander the Great, 163, 168, 189
Alexandria, 163, 164, 165, 168–70

Allen, Peter, 47
ammonia, 114, 115
ammonium, 28
Archimedes, 163
argon, 27
Arnold, Matthew, 159
Auchinleck, Claude, 163
Auden, WH, 54
Australia: and European settlement, 56–7; exploration of, 1–2; immigration, 59; interior of, 1, 2, 3, 53; legends of history, 2; railways in, 74–5, 78–81, 83–5; relations with the United Kingdom, 116, 192; and trade rules, 227; White Australia Policy, 58; *see also* global warming

Baker, Russell, 237

balloons (hot air), 2–3, 5; early, 5–6, 12–13, 255

Banting, William: *Letter on Corpulence*, 212

Barker, Pat: *Regeneration*, 158

Barlow, Dick, 37

Becks' Medical Jurisprudence (Beck and Beck), 235–7, 238, 239

Berlin, 45, 46

Biggles, 139, 152–3, 155

Blair, Tony, 100, 249

Blanchard, Jean Pierre, 6

Blue Max, The (film), 155

Boelcke, Max, 142

boilers; engine, 81–3, 86, 89–90, 180; hot water, 107

Brahe, Tycho, 49

Braidwood, 203

Braun, Eva, 41, 43, 154

Breatharians, the, 217–18

Brenner, Sydney, 36

Brisbane, 56–8, 78, 80–1, 94, 96, 99, 118, 119, 124, 139; and electrification, 133; firefighting in, 199–200

Brittain, Vera: *Testament of Youth*, 158

Bronowski, Joseph, 36

Brown, Arthur 'Roy', 149

Brown, Gordon, 249

Brown, Stan, 17, 24

bunker fire, 185–7, 194

Burke, Robert O'Hara, 1, 4–5, 11, 12

Burke and Wills expedition, 1–2, 4–5, 6, 7

burning/fire: and air pollution, 9, 34, 90–1, 255; and fireworks, 200–2; of fossil fuels, 9–10, 29, 34–5, 51–2, 255, 257; and 'preternatural combustion', 235–40, 242; and the World Trade Center, 204–7; and war, 151, 160; as warning beacons, 173–4; *see also* steam; sun(light)

Burnt By the Sun (film), 68

Bush, George H, 188

Bush, George W, 188, 248

Butterworth, George, 157

Byford, Charles, 141

Byford, Emma Eliza (nee Smith), 72, 78, 91, 113, 116

Byford, Frances, 91, 212–13

Byford, HL (Bert), 71, 72, 74, 78–9, 80, 91, 92, 113, 116, 133

Byford, Jack, 138

Byford, Linda, 60, 91, 93

calcium, 21, 26, 127

Calcraft, William, 72

Canberra, 108

capitalism: and tackling global warming, 69

capital punishment: as public spectacle, 72, 243

carbon, 16, 25–6, 30, 33, 126, 129, 259, 260, 262, 270, 271, 272 carbon dioxide, 11, 21, 26–7, 30, 34, 254

carbon monoxide, viii, 9, 21, 27, 34, 127

carbon taxes, 51, 109

carbon trading, 248, 270–1

Chaplin, Charlie, 41

Charles, Jacques, 6

chemistry, 18–23, 25–6; and life, 25–9, 51

Chifley, Ben, 85

Child, Julia, 211

China: and energy consumption, 260; and fireworks, 202; Great Wall, 173

chlorine, 22

Churchill, Winston, 41, 193, 194

Clark, William, 4

Clem, Bill, 231

climate change. *See* global warming

Clinton, Bill, 191

coal, 29, 33-5, 49, 79, 81, 85–6, 89, 95, 101, 108, 121, 126–7, 132, 178, 180-2, 185–6, 187, 190, 193–4, 237, 257, 261–2, 271

Cold War, the, 161

Coleridge, Samuel Taylor: *Rime of the Ancient Mariner*, 54–5

Concorde, the, 251–5

Cook, James, 92

Cooke, Alastair, 237

cooling: air-conditioners, 114, 118–19; ceiling fans, 120; and chlorofluorocarbons, 115–16; the Coolgardie safe, 112–13, 120; and energy use, 121; icemen and ice chests, 110–11, 112, 113, 115, 121, 122; and modern societies, 118–20; and perishable food, 113–14; refrigeration, 114–15, 116, 117–18

Courtenay, Bryce: *Potato Factory*, 132

Crick, Francis, 36

Davey, Humphrey, 126

da Vinci, Leonardo, 37

Day, Bob, 231

Day, Danny, 271

de Gaulle, Charles, 41

desalination, 269

Descartes, René, 168

developing world, 225–7, 264; and global warming, 260, 270; and resources, 259

Dewey, George, 178, 192

Dickens, Charles: *Bleak House*, 237–9, 242, 244; *The Old Curiosity Shop*, 125; *A Tale of Two Cities*, 125, 243

disease, 226–7, 264; bacterial, 16, 220–1; cancer, 58, 64–5, 68; diabetes, 28, 223 ; and food sources, 230–2; HIV/AIDS, 226; H5N1 bird flu, 220; yellow fever virus, 119

Doherty, Eric Chippendale, 60, 61, 91

Doherty, Finn, 197

Doherty, Helen (nee Chippendale), 124

Doherty, Jim, 197

Doherty, Penny (nee Stephens), 103, 116, 124, 238

Doherty, Peter: *The Beginner's Guide to Winning the Nobel Prize*, 118; education of, 15–18, 21, 23, 220; as research biologist, 93, 235; as veterinary scientist, 93, 118

Dr Zhivago (film), 73

Domagk, Gerhard, 20

Dulbecco, Renato, 36

Dupont, Jonas: *De Incendiis Corporis Humani Spontaneis*, 239

Durrell, Laurence: *The Alexandria Quartet*, 164

eagles and other birds, 38, 44, 47; symbolism of, 38–40, 46–7

economics: and consumption, 260; and inequality, 131, 209, 243–4, 259

Edinburgh, 99–104, 109, 117, 197, 203

Edison, Thomas, 126

education, 15–18

Edwards, Henry, 103

Egypt: Biblioteca Alexandrina, 162, 165–6, 168; Great Library of Alexandria, 162–3, 167, 174

Ehrlich, Paul, 20

Einstein, Albert, 50

Eisenhower, Dwight D, 247

energy generation, 256; and carbon taxes, 109, 270, 271; and fossil fuels, 9–10, 29–31, 33–5, 49, 51–2, 70, 85, 115, 121, 127, 133, 180, 184–5, 193, 252, 253, 271; and health, 9–10, 101, 180, 256; and industrial and technological developments, 257–8; and methane, 30–1, 32, 126–7, 259; nuclear power, 95, 121, 262; renewable energy, 35, 50, 51–2, 108–9, 121, 262–3, 269; and sunlight, 25, 28, 171; and unscrupulous suppliers, 106, 108

energy supplies, 194; and war, 187–9

Enlightenment, the, 13

environment: change in, 172; degradation of, 34, 234, 259, 261, 264, 268–9, 271, 273; and sustainability, 229–30, 232

ethanol, 31–2, 56

Euclid, 163

Everett, Edward, 246
exploration, 132

farming: and energy needs, 32, 33
Faulks, Sebastian, 239; *Birdsong*, 158
Fawkes, Guy, 200–1
Feynman, Richard, 273
firefighting, 197–205, 209
Flannery, Tim: *The Weather Makers*, 272
Fokker, Anton, 143, 149
Fokker Dreidekker, 140, 143, 146, 147, 150
Fokker Eindekker, 145
food and diet, 210–18, 227–8; and animals, 220–1; and fish stocks, 228–32; and hunter–gatherers, 224–5; and obesity, 222–3; and religion, 218–20, 221–22; and transition from transitional lifestyles, 223–4; and underdevelopment, 225–6
Forster, EM, 58
fossil hydrocarbons, 10, 22, 29, 33, 34, 52, 121, 193, 262; wars *see* United States
France, 122; under Louis XIV, 69–70, 243; French Revolution, 70, 125, 243
Franklin, Benjamin, 34, 39, 46
Franklin, Sir John, 268
Fraser, Malcolm, 247

Freeman, Walter, 111–12
Freon, 116
Fresnel, Augustin, 170
Fried Green Tomatoes (film), 227

Gaslight (film), 127
Germany: anti-Semitism in, 155, 161; economic collapse in, 42, 44; German Navy, 192, 194; Nazi period, 20, 40–4, 45–6, 144, 153, 161, 167; science in, 50
Gesner, Abraham, 30
Gibbons, Lewis Grassic (JL Mitchell): *Sunset Song*, 158
Gingrich, Newt, 246
global cooling, 53
globalisation, 76, 164, 257; and economic inequality, 131, 209; and opportunity, 260
global warming (man-made), 13, 49, 55, 68, 70, 230, 234, 242; and acid sunscreen idea, 52–3; and atmospheric pollution, 10, 260–1; and behavioural change, 123; effect on Africa, 13–14, 55; effect on Australia, 8, 13, 55, 261, 267, 271; effect on the United States, 8, 55, 172–3, 261, 268; and emission reductions, 69, 270–2; and greenhouse gases, 26, 30, 52, 116, 121, 127, 249, 259, 260–1; and the International Panel for Climate Change

(IPCC), 261, 264, 266, 267, 268; and politics and government, 248–9, 265, 270; and population, 121; and renewable energy, 35, 50, 51–2, 69, 121, 122, 262–3; and rising sea levels, 172–3, 250, 264, 268; and 'tipping points', 263; and weather patterns and temperature, 8, 55, 122, 123, 173, 250, 266–7; *see also* energy generation

glucose, 28, 63, 214, 216, 223

Goebbels, Joseph, 44, 154

Goering, Hermann, 142, 144, 149, 154, 160

gold (aurum), 22

Golstein, Joe, 213

Gore, Al, 249, 270; *An Inconvenient Truth*, 263–4

Grand Illusion (film; Renoir), 156

Grant, Ulysses S, 75

Greece: Athenian democracy, 163

Gridley, Charles, 192

Grube, Fritz Lehmann, 37

Guiliano, Mireille: *French Women Don't Get Fat*, 211

Guillemin, Roger, 36

Halberstadt, 142, 146

Hanson, Christine Lee, 207

Hanson, Peter Burton, 207

Hanson, Sue Kim, 207

Harvey, Sir Eliab, 177

Haslam, Garth: *Anomalies*, 239

Hauser, Gayelord: *Diet Does It*, 212, 213, 225

heating (domestic), 97–102; and building design, 97–8, 109, 121; and energy supplies, 106–8; kerosene, 135; natural gas, 97, 127; and pollution, 101; and the poor, 102–4, 105–6, 135; in Roman England, 104–5

helium, 7, 50

Hell's Angels (film), 148, 154–5

Henry VIII, 176, 218

Hepburn, Audrey, 211

Himmler, Heinrich, 40, 41

Hitler, Adolf, 40–2, 43, 154, 161, 189; *Mein Kampf*, 41

HMAS *Sydney*, 88, 138

HMS *Endymion*, 177

HMS *Prince of Wales*, 192

HMS *Repulse*, 192

HMS *Temeraire*, 176–8

HMS *Victory*, 176, 177

HMS *Warrior*, 87, 178

Homer: *Odyssey*, 256

Howard, John, 248

Hughes, Howard, 148, 154

humans, 53, 258; and body cells, 66–8; and consciousness, 11; human spirit, 11, 12, 13, 167; and population, 14, 121, 258–9, 260; importance of reading and writing, 135; skin,

62, 65–8; and temperature, 62–4; and uncertainty, 208

Hume, David, 102

Hurley, Frank, 132

Hussain, Saddam, 188

hydrocarbons, 52, 187, 190, 195

hydrogen, 6–7, 12–13, 22, 26, 29–30, 32, 50, 262, 272

hydrogen sulphide, 16

Immelman, Max, 142, 145

industrial revolution, 34, 71, 127, 178, 257

Iolanthe (Gilbert and Sullivan), 222

Iraq War, 100, 161, 183, 195, 227, 248

Ireland, 174

iron, 19, 20, 22

Islam, 164–5; intellectual flowering of, 168; and modernism, 165; and science, 165–6

Israeli–Palestinian conflict, 189

Jefferson, Thomas, 112

Jeffries, John, 6

John the Baptist, Saint, 218

Johns, WE, 152–3

Johnson, Lyndon B, 190

Johnson, Samuel, 210

Jones, Casey, 78, 80

Judge, Mychal, 205

Kahn, Louis, 36

Kelly, Ned, 2, 11

Keneally, Tom: *The Great Shame*, 132

Kennedy, John F, 247

Kennedy, Rosemary, 112

Kepler, Johannes, 49

King, James, 2

Knox, John, 103

Kormoran, 138

Kougionas, Vassily, 271

Latham, Mark, 244

Lavoisier, Antoine, 125–6

Lawrence, TE, 194

Lawson, Nigella, 211

Le Cat, Nicholas, 236, 238, 239

Levi, Primo, 18, 20–1; *The Periodic Table*, 19, 24

Lawrence, DH, 156–7

Lawrence, Frieda (nee von Richthofen), 156–7

Lewis, Meriwether, 4

lighting: and air pollution, 128, 134, 136; candles, 124–5, 127, 128–9, 131; and coal gas, 132; and electricity, 126, 135, 171; and energy use, 126; gaslight, 126, 127; and kerosene, 132, 133–4, 135, 171; lanterns, 133, 134, 135; lighthouses, 168–9, 170–1; and mantles, 134; oil (pressure) lamps, 127–8, 129–30, 134–5; and paraffin wax, 129; and reading and

writing, 128–9, 135; and
     tallow, 128–9; and whale oil,
     130, 131, 171
Lillienthal, Otto, 38
Lincoln, Abraham, 228, 246
Lindbergh, Charles, 157
lobotomies, 112
London, 128, 177; and fire,
     202, 203
Lost Squadron, The (film), 156
Louis XIV, 69–70, 243, 258
Louis XV, 69, 70
Louis XVI, 5, 243, 255
Lovelock, James, 52, 53
Lowe, Ian: *A Big Fix*, 272

McCain, John, 249
McCall Smith, Alexander:
     Expresso Tales, 100; *44 Scotland
     Street*, 100
McCormick, AP, 112
McCourt, Frank: *Angela's Ashes*,
     103
McKellar, Dorothea, 55
Mahfouz, Naguib: *Cairo Trilogy*,
     164
Marie Antoinette (Queen), 6,
     243, 255
Mary Queen of Scots, 103
Maxim machine gun, 149–50
media, 245, 248; and global
     warming, 265–6
Melbourne, 1, 59, 267
Melville, Herman: *Moby Dick*,
     130, 131

Memphis, 75, 97, 106–7, 108,
     119–20, 199
Mendeleev, Dmitri, 17
methanol, 31–2, 61, 141
Middle East, 100, 161, 183,
     188, 194–5
Miller, Roger, 77
Millet, Nicole, 236, 239
mining: and gases, 127
Montgolfier brothers, 5–6, 11
Montgomery, Bernard Law,
     145, 164
Mubarak, Suzanne, 165
Murder on the Orient Express
     (film), 83
Murdoch, Rupert, 248
Mussolini, Benito, 18, 41

Nature, 264
Nelson, Horatio, 176, 177
Newton, Isaac, 50
New York, 204–6
nitrogen, 11, 21, 26–9, 121,
     254; liquid, 118
Nolan, Sidney, 2, 12, 53
Nurse, Paul, 37

O'Doherty, Kevin Izod, 57
O'Neill, Eugene: *The Iceman
     Cometh*, 110–11, 123
oil, 25, 29–30, 32–4, 47, 85, 89,
     108, 121, 127, 171, 188–9,
     194–5, 261–2, 271; lamps *see*
     lighting
Owen, Wilfred, 157

oxygen, 27, 29, 33, 43, 64, 125–6, 206, 213, 214, 253, 254, 255

Palmerston North, 269
Peanuts, 143–4
Pepys, Samuel, 128–9
Philadelphia, 202
phosphorus, 21
Planck, Max, 50
pollution: atmospheric, 9–10, 13, 91–2, 101, 260–1, 273
Popkin, Cedric, 149, 150
potassium, 22
Power, Annie, 15, 16–17, 23
Pritchard, EW, 72
Ptolemy I, 162
Ptolemy III, 163

Rankin, Ian, 103
recreation, 232, 233, 234, 242–3
Reid, Hugh, 103
religion, 218–19
Renoir, Jean, 156
Rickover, Hyman B: *How the Battleship* Maine *Was Destroyed*, 184
Riefenstahl, Leni: *The Triumph of the Will*, 42
Rittenhouse, David, 34
RMS *Lusitania*, 88, 189–90
RMS *Mauretania*, 88, 194
RMS *Titanic*, 88, 179–81, 185–7
Roehm, Ernst, 153
Rolls-Royce Merlin engine, 150

Rommel, Erwin, 154, 163
Roosevelt, Franklin D, 41, 188, 193, 247
Roosevelt, Theodore, 183, 191, 246
Rotary engine, 140, 146–7
Roth, Philip: *The Plot Against America*, 157

Sassoon, Siegfried, 158
Schulz, Charles, 143, 144, 148
Schwarzenegger, Arnold, 248
Science, 264, 271
science: and the dark ages, 167–8; discoveries in, 20, 125, 126, 170, 213; education, 17–19, 20, 23–4; and energy, 50, 121; and evolution, 233; ignorance of, 22; and other disciplines, 222; scientific method, 92, 267; *see also* chemistry; scientific research
scientific institutes and societies, 50; Crown Research Institutes (NZ), 269; International Laboratory for Research in Animal Diseases (Nairobi), 251; Royal Society of London, 129; St Jude Children's Research Hospital, 117, 198–9; Salk Institute, 36–7

scientific research, 222, 258; biomedical, 117; and fire risk, 197–8; and food sources, 230–1; and improvements in health, 258; and refrigeration, 117–18

Scotland, 102

Seregeldin, Ismael, 162, 165

Shackelton, Ernest: *South*, 132

Shadyac, Dick, 199

Shakespeare, William, 53–4

Sherman, Alan, 62

silver (argentum), 22

Simpson, Stuart, 141

singing: railway songs, 77

Smith, Adam, 102

Smith, Mike, 213

Smith, Peter Bedford, 72

SMS *Emden*, 138

Snoopy, 144, 148, 151

Snow, CP, 23

sodium, 22, 113

Sopwith Camel biplane, 144, 145–7, 149, 150, 151, 152

Soviet Union, 68

Spandau (Maxim) machine gun, 149

Spanish Civil War, 159

species depletion, 273

Stalin, Joseph, 41, 68

steam: and accidents, 89–90, 93–4, 184–7, 190, 194, 195; and cooking, 92–3; and machinery, 71, 87, 94, 95; and powerhouses, 82, 88–9, 95; and railways, 71–3, 74, 81–7, 89, 90–5; steam cars, 82–3; steamships, 87–8, 90, 95, 179–81, 184–6, 194 ; and tea-making, 95–6

steam turbine, 31, 87, 88, 115, 194

Stephens, Bill, 235

Stoker, Bram, 141

Stoker, Dacre, 141

Stuart, John McDougall, 2

sulphur, 18–19, 52

Sultana, 90

sun(light), 48–50; and the arts, 53–5; and energy, 25, 28, 50–1; and global warming, 49, 55, 68; ultra-violet radiation and sunburn, 56, 57–62, 64–8

Swan, Joseph, 126

Sydney, 119

Sykes–Picot agreement, 194

technology, 94; and death and destruction, 182–3, 206; technological innovation, 123, 180, 195, 229, 248, 256–7, 271

terrorism, 164, 174, 191, 195, 204–8; and fundamentalism, 166–7, 219

thermodynamics, first law of, 25

Theroux, Paul: *Happy Isles of Oceania*, 223

Tickle, Reg, 17, 24

totalitarianism, 44–5, 154, 195; and rhetoric, 42, 43, 44; and security, 46; and tackling global warming, 69

transportation: air travel, 7, 13, 37–8, 42–3, 170, 180, 207–8, 225, 229, 251–5, 256; bicycles, 270; railways, 71–87, 89, 90–2; road vehicles, 30–3, 83, 91, 256, 263; ships, 87–8, 169, 171, 179–82, 185–6, 193–4, 228; *see also* war

triple expansion engine, 87–88

Trotsky, Leon, 111

Trudeau, Gary, 144

Truman, Harry S, 247

Turner, JMW, 53; *The Fighting Temeraire*, 176, 177–8

Twain, Mark: *The Adventures of Huckleberry Finn*, 228

Udet, Ernst, 153–4

United Kingdom, 129; entry into the EEC, 116, 270; Royal Navy, 176–8, 192–3, 194; railways in, 71–4

United States: American South, 58, 120; Atlantic coast, 169–72; Civil War, 5, 75, 159, 171, 183, 246; and energy consumption, 260; exploration of, 4; 'fossil hydrocarbon wars' of, 187–8; founding of, 39; and the 'free world', 247–8; Independence Day, 201; Mississippi delta, 227, 228; railways in, 75–7, 80, 85; US Navy, 178–9, 182, 183–5, 187, 190–2; *see also* global warming

USS *Arizona*, 189

USS *C. Turner Joy*, 190

USS *Cole*, 176, 191, 193

USS *Maddox*, 190

USS *Maine*, 182, 191

USS *Olympia*, 87, 179, 182, 191

USS *Oregon*, 185

Vaxjo, 270

Verne, Jules: *Around the World in 80 Days*, 256

Vernon, Tom: *Fat Man on a Bicycle*, 211

Vickers (Maxim) machine gun, 150

Victoria (Queen), 193

Vietnam War, 190

Vonnegut, Kurt Jr: *God Bless You, Mr Rosewater*, 203; *Slaughterhouse Five*, 43

von Richthofen, Lothar, 153

von Richthofen, Manfred (the Red Baron), 137, 141, 143, 144, 145, 149, 151, 153, 154, 155, 156, 157, 160, 255

Von Ryan's Express (film), 79

von Stauffenberg, Claus, 154

von Stroheim, Erich, 156

Voss, Werner, 155

Walton, Isaac: *The Compleat Angler*, 232

war: aircraft in, 139–40, 141,

142–3, 145–51, 154;
Australian War Memorial,
141; and bombed and burning
cities, 159; books about,
140–2, 145, 151; Imperial War
Museum, 145, 146; and naval
forces, 176–9, 182, 185, 187,
189–93; over land and
resources, 187–9; and solving
problems, 196; technology in,
150, 160; *see also* World War I;
World War II
weather patterns. *See* global
warming
Webb, Sim, 78
whaling, 130–3
Wilhelm II, 192
Wills, William John, 2, 4–5; 11,
12
Wilson, Woodrow, 190
Wolfe, Tom, 12

World War I, 137, 142, 143, 144,
160, 194–5; Australia and,
138, 141; literary legacy of,
158; slaughter in, 141, 157,
159, 161, 182; and the United
States, 190; *see also* war
World War II, 41, 42–3, 137,
144, 169; and Australia,
138–9, 141; battles and
campaigns, 138–9, 149, 163–4,
189; bombing raids in, 43, 44,
133, 139; and fuel shortages,
134; and Japan, 188, 192; Nazi
extermination camps, 20, 44,
46, 160; and the United States,
39
Worrals, 152
Wright brothers, 170, 255
Wyn, Hedd, 157

zinc, 6, 60